Monitoring, Simulation, Prevention and Remediation of Dense and Debris Flows III

WIT*PRESS*

WIT Press publishes leading books in Science and Technology.
Visit our website for the current list of titles.
www.witpress.com

WIT*eLibrary*

Home of the Transactions of the Wessex Institute.
Papers presented at Debris Flow III are archived in the WIT eLibrary in volume 67 of WIT Transactions on Engineering Sciences (ISSN 1743-3533).
The WIT eLibrary provides the international scientific community with immediate and permanent access to individual papers presented at WIT conferences.
http://library.witpress.com

THIRD INTERNATIONAL CONFERENCE ON
MONITORING, SIMULATION, PREVENTION AND REMEDIATION
OF DENSE AND DEBRIS FLOWS

DEBRIS FLOW III

CONFERENCE CHAIRMEN

D. de Wrachien
University of Milan, Italy

C.A. Brebbia
Wessex Institute of Technology, UK

INTERNATIONAL SCIENTIFIC ADVISORY COMMITTEE

Organised by
University of Milano, Italy
Wessex Institute of Technology, UK

Sponsored by
WIT Transactions on Engineering Sciences
EurAgEng: European Society of Agricultural Engineers
CIGR: International Commission of Agricultural Engineering

Supported by
The Lombardy Region, Italy

WIT Transactions

Transactions Editor

Carlos Brebbia
Wessex Institute of Technology
Ashurst Lodge, Ashurst
Southampton SO40 7AA, UK
Email: carlos@wessex.ac.uk

Editorial Board

Monitoring, Simulation, Prevention and Remediation of Dense Debris Flows III

Editors

D. de Wrachien
State University of Milan, Italy

C.A. Brebbia
Wessex Institute of Technology, UK

WIT*PRESS* Southampton, Boston

D. de Wrachien
State University of Milan, Italy

C.A. Brebbia
Wessex Institute of Technology, UK

Published by

WIT Press
Ashurst Lodge, Ashurst, Southampton, SO40 7AA, UK
Tel: 44 (0) 238 029 3223; Fax: 44 (0) 238 029 2853
E-Mail: witpress@witpress.com
http://www.witpress.com

For USA, Canada and Mexico

Computational Mechanics Inc
25 Bridge Street, Billerica, MA 01821, USA
Tel: 978 667 5841; Fax: 978 667 7582
E-Mail: infousa@witpress.com
http://www.witpress.com

British Library Cataloguing-in-Publication Data

A Catalogue record for this book is available
from the British Library

ISBN: 978-1-84564-442-0
ISSN: 1746-4471 (print)
ISSN: 1743-3533 (online)

The texts of the papers in this volume were set individually by the authors or under their supervision. Only minor corrections to the text may have been carried out by the publisher.

Printed in Great Britain by Martins the Printers.

Preface

This book contains papers presented at the Third International Conference on Debris Flow including all aspects of Debris Flow Monitoring, Modelling, Hazard Assessment, Mitigation Measures, Extreme Events, Erosion, Slope Instability and Sediment Transport, held in Milano, Italy, in 2010. The Conference was jointly organised by the State University of Milano and the Wessex Institute of Technology, UK, with the co-sponsorship of EurAgEng (European Society of Agricultural Engineers) and CIGR (International Commission of Agricultural Engineering) and the support of the Lombardy Region, Italy. This successful series of Conferences first started in Rhodes, Greece (2006) and continued in New Forest, UK (2008).

Debris and hyper-concentrated flows are among the most destructive of all water related disasters. They affect both rural and urban areas in a wide range of morpho-climatic environments, and in recent years have attracted more and more attention from the scientific and professional communities and concern from the public due to the death toll they claim.

The increased frequency of these natural hazards, coupled with climatic change predictions and urban development, suggests that they are set to worsen in the future.

The Conference brought together engineers, scientists and managers from across the globe to discuss the latest scientific advances in the field of dense and hyper-concentrated flows, as well as to improve models, assess risk, develop hazard maps based on model results and to design prevention and mitigation measures.

The book contains Sections on the following topics:

- Debris Flow Modelling
- Debris Flow Triggering
- Risk Assessment and Hazard Mitigation
- Sediment Transport and Debris Flow Monitoring & Analysis

The Editors would like to thank all the Authors for their excellent contributions as wells as the members of the International Scientific Advisory Committee for their help in reviewing both the abstracts and the papers included in this book. The quality of the material makes this volume a most valuable and up-to-date tool for professionals, scientists and managers to appreciate the state-of-the-art in this important field of knowledge

The Editors
Milano, 2010

Contents

Section 1: Debris flow modelling

Mechanical and fluid-dynamic behaviour of debris and hyper-concentrated flows: overview and challenges
D. De Wrachien, S. Mambretti & C. Deangeli 3

One-dimensional finite volume simulation of real debris flow events
L. Schippa & S. Pavan 17

Debris flow modelling accounting for large boulder transport
C. Martinez, F. Miralles-Wilhelm & R. Garcia-Martinez 29

New formulas for the motion resistance of debris flows
D. Berzi, J. T. Jenkins & E. Larcan 41

Rheological behaviour of pyroclastic debris flow
A. M. Pellegrino, A. Scotto di Santolo, A. Evangelista & P. Coussot 51

Section 2: Debris flow triggering

The triggered mechanism of typhoon-induced debris flows and landslides over mainland China
G. P. Zhang, J. Xu, F. W. Xu, L. N. Zhao, Y. M. Li, J. Li, X. D. Yang & J. Y. Di 65

Debris flow occurrences in Rio dos Cedros, Southern Brazil: meteorological and geomorphic aspects
M. Kobiyama, R. F. Goerl, G. P. Corrêa & G. P. Michel 77

Soil moisture retrieval with remote sensing images for debris flow forecast in humid regions
Y. Zhao, H. Yang & F. Wei 89

Debris flow induced by glacial lake break in southeast Tibet
Z. L. Cheng, J. J. Liu & J. K. Liu 101

Experience with treatment of road structure landslides by innovative methods of deep drainage
O. Mrvík & S. Bomont 113

Technical protection measures against natural hazards taken by the Austrian Federal Service for Torrent, Erosion and Avalanche Control
F. J. Riedl 125

Section 3: Risk assessment and hazard mitigation

The distribution of debris flows and debris flow hazards in southeast China
F. Wei, Y. Jiang, Y. Zhao, A. Xu & J. S. Gardner 137

Evaluation of sediment yield from valley slopes: a case study
F. Ballio, D. Brambilla, E. Giorgetti, L. Longoni, M. Papini & A. Radice 149

Shallow landslide full-scale experiments in combination with testing of a flexible barrier
L. Bugnion & C. Wendeler 161

Landslide in a catchment area of a torrent and the consequences for the technical mitigation concept
F. J. Riedl 175

Regional methods for shallow landslide hazard evaluation: a comparison between Italy and Central America
D. Brambilla, L. Longoni & M. Papini 185

Section 4: Sediment transport and debris flow monitoring and analysis Special session organised by Daniele De Wrachien, Gian Battista Bischetti, Francesco Gentile & Luca Mao

Erosion and sediment transport modelling in Northern Puglia watersheds
F. Gentile, T. Bisantino & G. Trisorio Liuzzi 199

Restoration of a degraded torrential stream by means of a flood control system: the case of Arroyo del Partido stream (Spain)
J. A. Mintegui Aguirre, J. C. Robredo Sánchez, C. De Gonzalo Aranoa & P. Huelin Rueda 213

The effects of large wood elements during an extreme flood in a small tropical basin of Costa Rica
L. Mao & F. Comiti 225

Rheological properties and debris-flow modeling in a southern Italy watershed
T. Bisantino, P. Fischer, F. Gentile & G. Trisorio Liuzzi 237

Formation, expansion and restoration of a sedimentation fan: the case of the Arroyo del Partido stream (Spain)
J. A. Mintegui Aguirre, J. C. Robredo Sánchez, L. Mao & M. A. Lenzi 249

Dynamics of changes of bed load outflow from a small glacial catchment (West Spitsbergen)
W. Kociuba, G. Janicki & K. Siwek 261

Author Index 271

Section 1
Debris flow modelling

Mechanical and fluid-dynamic behaviour of debris and hyper-concentrated flows: overview and challenges

D. De Wrachien[1], S. Mambretti[2] & C. Deangeli[3]
[1]*Department of Agricultural Engineering, State University of Milan, Italy*
[2]*DIIAR, Politecnico di Milano, Italy*
[3]*DITAG, Politecnico di Torino, Italy*

Abstract

Debris and hyper-concentrated flows are among the most destructive of all water-related disasters. They mainly affect mountain areas in a wide range of morphoclimatic environments and in recent years have attracted more and more attention from the scientific and professional communities and concern from public awareness, due to the increasing frequency with which they occur and the death toll they claim. In this context, achieving a set of debris and hyper-concentrated flow constitutive equations is a task that has been given particular attention by scientists during the second half of the last century.

In relation to these issues, this paper reviews the most updated and effective geotechnical and fluid-dynamic procedures nowadays available, suitable to predict the triggering and mobilising processes of these phenomena, and proposes a mathematical model that is able to assess the depth of the wave and the velocities of the liquid and solid phases of both non-stratified (mature) and stratified (immature) flows following flash-floods and dam-break events in one and two dimensional cases.

Different experimental cases of dam-break situations in a square section channel were considered for the purpose of comparing results.

These tools will allow, on one hand, to better focus on what to observe in the field and, on the other hand, to improve both mitigation measures and hazard mapping procedures.

Keywords: debris flow, rheological behaviour of the mixture, slope failure, numerical models, laboratory and field tests.

1 Introduction

Debris and hyper-concentrated flows are among the most destructive of all water-related disasters. They mainly affect mountain areas in a wide range of morphoclimatic environments and in recent years have attracted more and more attention from the scientific and professional communities and concern from public awareness due to the increasing frequency with which they occur and the death toll they claim. These phenomena do not allow a sufficient early warning, as they are characterised by a very short time-scale and, therefore, defence measures should be provided, especially when they are associated with flash floods or dam failures. To this end, the identification of effective procedures aimed at evaluating the probability of these extreme events and the triggering and mobilising mechanism has become an essential component of the water and land use planning processes. This concept leads to a new integrated risk management approach, which comprises administrative decisions, organisation, operational skill and the ability to implement suitable policies. The broadness of the question requires approaches from various perspectives.

To this end, the dynamic behaviour of these hyper-concentrated water sediment mixtures and the constitutive laws that govern them plays a role of paramount importance.

Debris flow modelling requires a rheological pattern (or constitutive equation) that provides an adequate description of these flows.

One of the main difficulties met by the approaches available is linked to their validation either in the field or in a laboratory environment. Greater research needs to be directed towards a thorough investigation of the above mentioned issues.

Such knowledge is essential in order to assess the potential frequency of these natural hazards and the related prevention and mitigation measures.

With reference to these issues, this paper aims to provide the state-of-the-art of debris flow rheology, modelling and laboratory and field investigation, along with a glance to the direction that debris flow in-depth studies are likely to follow in future.

2 Debris flow model development

A thorough understanding of the mechanism triggering and mobilising debris flow phenomena plays a role of paramount importance for designing suitable prevention and mitigation measures. Achieving a set of debris flow constitutive equations is a task which has been given particular attention by the scientific community (Julien and O'Brien [33]; Chen [9]; Takahashi [35]). To properly tackle this problem relevant theoretical and experimental studies have been carried out during the second half of the last century.

Research work on theoretical studies has traditionally specialised in different mathematical models. They can be roughly categorized on the basis of three characteristics: the presence of bed evolution equation, the number of phases and the rheological model applied to the flowing mixture (Ghilardi et al. [24]).

Most models are based on the conservation of mass and momentum of the flow, but only a few of them take into account erosion/deposition processes affecting the temporal evolution of the channel bed.

Debris flows are mixtures of water and clastic material with high coarse particle contents, in which collisions between particles and dispersive stresses are the dominant mechanisms in energy dissipation.

The rheological property of a debris flow depends on a variety of factors, such as suspended solid concentration, cohesive property, particle size distribution, particle shape, grain friction and pore pressure.

Various researchers have developed models of debris flow rheology These models can be classified as: Newtonian models (Johnson [32]), linear and non linear viscoplastic models (O'Brien et al. [41]), dilatant fluid models (Bagnold [4]), dispersive or turbulent stress models (Arai and Takahashi [2]), biviscous modified Bingham model (Dent and Lang [15]), and frictional models (Norem et al. [40]). Among these, linear (Bingham) or non-linear (Herschel-Bulkey) viscoplastic models are widely used to describe the rheology of laminar debris/mud flows (Jan, 1997).

Because a debris flow, essentially, constitutes a multiphase system, any attempt at modelling this phenomenon that assumes, as a simplified hypothesis, homogeneous mass and constant density, conceals the interactions between the phases and prevents the possibility of investigating further mechanisms such as the effect of sediment separation (grading).

Modelling the fluid as a two-phase mixture overcomes most of the limitations mentioned above and allows for a wider choice of rheological models such as: Bagnold's dilatant fluid hypothesis (Takahashi and Nakagawa [56]), Chezy type equation with constant value of the friction coefficient (Hirano et al. [27]), models with cohesive yield stress (Honda and Egashira [28]) and the generalized viscoplastic fluid Chen's model (Chen and Ling [10]).

Notwithstanding all these efforts, some phenomenological aspects of debris flow have not been understood yet, and something new has to be added to the description of the process to reach a better assessment of the events. In this contest, the mechanism of dam-break wave should be further investigated. So far, this aspect has been analysed by means of the single-phase propagation theory for clear water, introducing in the De Saint Venant (SV) equations a dissipation term to consider fluid rheology (Coussot [12]; Fread and Jin [23]).

Many other models, the so-called quasi-two-phase-models, use SV equations together with erosion/deposition and mass conservation equations for the solid phase, and take into account mixture of varying concentrations. All these models feature monotonic velocity profiles that, generally, do not agree with experimental and field data.

2.1 Rheology

The rheological property of debris and hyper-concentrated flows depends on a variety of factors, such as the suspended solid concentration, cohesive property, size distribution, particle shape, grain friction, and pore pressure. So, modelling these flows requires a rheological model (or constitutive equation) for sediment-water mixtures.

A general model which can realistically describe the rheological properties of debris flow should possess three main features (Chen [9]). The model should:

- describe the dilatancy of sediment-water mixtures;
- take into account the so-called soil yield criterion, as proposed by Mohr-Coulomb;
- assess the role of intergranular or interstitial fluid.

The earliest of such rheological models was empirically formulated by Bagnold [4].

On the whole, a rheological model of debris and hyper-concentrated flows should involve the interaction of several physical processes. The non-Newtonian behaviour of the fluid matrix is ruled, in part, by the cohesion between fine sediment particles. This cohesion contributes to the yield stress, which must be exceeded by an applied stress in order to initiate fluid motion.

In view of theoretical soundness behind the development of different non-Newtonian fluid models, Bailard [5] and Hanes [25] have questioned the validity of Bagnold's empirical relations. Limitations in Bagnold's model may be attributed to the ambiguity in the definition of some rheological characteristics as the grain stresses.

To overcome these problems, Chen [9] developed a new generalised viscoplastic fluid (GVF) model, based on two major rheological properties (i.e. the normal stress effect and soil yield criterion) for general use in debris flow modelling.

The analysis Chen conducted on the various flow regime of a granular mixture identified three regimes: a quasi-static one, which is a condition of incipient movement with plastic behaviour, a microviscous one at low shear rates, in which viscosity determines the mixture behaviour, and finally a granular inertial state, typical of rapid flowing granular mixtures, dominated by intergranular interactions.

All the models previously reviewed feature monotonic velocity profiles that, generally, do not agree with experimental and field data. In many tests (Takahashi [53]) "*S*" reversed shaped trends have been observed, where the maximum shear rate is not achieved near the bed, but rather between the bed and the free surface. The main discrepancy is derived from the assumption of a debris flow as a uniform mixture. In fact, the solid concentration distribution is usually non-uniform due to the action of gravity, so that the lower layer could, consequently, have a higher concentration than the upper layer. Higher concentration means higher cohesion, friction and viscosity in the flow.

Wan [58] proposed a multilayered model known as the laminated layers model that features a stratified debris flow into three regions from the bed to the surface: a bed layer, in which an additional shear stress is dominant in momentum exchange; an inertial layer, where the dispersive stress of the grains is dominant; and an upper viscoplastic layer, which can be represented by the Bingham's model.

The one-layer models are unable to adequately feature the entire thickness of the flow and, therefore, it has recently become common to use multi-layers models that combine two or more constitutive relationships in order to analyse adequately these phenomena. The coefficients of the rheological models have wide ranges of variation and, therefore, in evaluating them considerable errors are committed. On the other hand, some empirical equations of velocity are necessary in any debris flow disaster-forecasting measure, although the hydraulics of debris-flow is not theoretically comparable to that of a traditional water flow.

2.2 Triggering and mobilising processes

Debris flow resulting from flash flood or a sudden collapse of a dam (dam-break) are often characterised by the formation of shock waves caused by many factors such as valley contractions, irregular bed slope and non-zero tailwater depth. It is commonly accepted that a mathematical description of these phenomena can be accomplished by means of 1D SV equations (Bellos and Sakkas [6]).

During the last Century, much effort has been devoted to the numerical solution of the SV equations, mainly driven by the need for accurate and efficient solvers for the discontinuities in dam-break problems.

A rather simple form of the dam failure problem in a dry channel was first solved by Ritter [46] who used the SV equations in the characteristic form, under the hypothesis of instantaneous failure in a horizontal rectangular channel without bed resistance. Later on, Stoker [50], on the basis of the work of Courant and Friedrichs [11], extended the Ritter solution to the case of wet downstream channel. Dressler [19] used a perturbation procedure to obtain a first-order correction for resistance effects to represent submerging waves in a roughing bed.

Lax and Wendroff [35] pioneered the use of numerical methods to calculate the hyperbolic conservation laws. McCormack [39] introduced a simpler version of the Lax-Wendroff scheme, which has been widely used in aerodynamics problems. Van Leer [57] extended the Godunov scheme to second-order accuracy by following the Monotonic Upstream Schemes for Conservation Laws (MUSCL) approach. Chen [7] applied the method of characteristics, including bed resistance effects, to solve dam-break problems for reservoir of finite length.

Sakkas and Strelkoff [47] provided the extension of the method of the characteristics to a power-law cross section and applied this method to a dam break on a dry right channel in the case of rectangular and parabolic cross section shapes. Strelkoff and Falvey [52] presented a critical review of numerical methods of characteristics of power-law cross sections.

Hunt [29] proposed a kinematic wave approximation for dam failure in a dry sloping channel.

Total Variation Diminishing (TVD) and Essentially Non Oscillation (ENO) schemes were introduced by Harten [26] for efficiently solving 1D gas dynamic problems. Their main property is that they are second order accurate and oscillation free across discontinuities.

Recently, several 1D and 2D models using approximate Riemann solvers have been reported in the literature. Such models have been found very successful in solving open channel flow and dam-break problems.

In the past ten years, further numerical methods to solve flood routing and dam-break problems, have been developed that include the use of finite elements or discrete/distinct element methods (Asmar et al. [3]).

Finite Element Methods (FEMs) have certain advantages over finite different methods, mainly in relation to the flexibility of the grid network that can be employed, especially in 2D flow problems.

Mambretti et al. [38] and De Wrachien and Mambretti [17, 18] used an improved TVD-Mc Cormack-Jameson scheme to predict the dynamics of both mature (non-stratified) and immature debris flow in different dam break conditions.

3 Laboratory and field studies

To validate both the rheological and dynamic models, herewith described, comparisons need to be made between their predictions and results of laboratory and field tests. Agreements between the computational and experimental results are essential since they allow the assessment of the models' performance and suggest feasible development of the research.

The experimental point of view in debris flow research, however, encounters considerable problems that are yet to be fully overcome, connected largely to the accuracy of measuring techniques and flow simulation in experimental tests. Lastly, field studies are probably the most difficult and costly study approach of debris flow; the difficulties encountered are connected to their considerable complexity and the difficulty of direct observation. The exceptional and infrequent conditions in which debris flows occur do not generally permit a sufficient number of observations for the same type of field reality to deduce the specific behavioural laws for that area. Reference to different territorial situations also highlights another problem: that of the homogeneity of data, given the substantial territorial peculiarity in which the phenomena occur. Besides, field data are essential in determining the quality of any mathematical model, as they are especially important for estimating velocity, discharge, concentration, yield stress, viscosity and grain-size.

This need requires the use of laboratory experimentation when the previous problems cannot be overcome, and in certain cases it is the only possible path to follow.

Within this ground, many experiments have been carried out, ranging from solid transport (little amount of particles in a large environment of clear water) to dry granular flow, where water is not present.

An empirical picture of debris flow physics can be drawn from a combination of real-time field observations (Okuda et al. [42]); detailed measurements during controlled field and laboratory experiments (Takahashi [54]), and analyses of debris flow paths and deposits (Fink et al. [21]).

Few reliable techniques exist to measure properties of flowing debris. Grossly invasive procedures such as plunging buckets or sensors into the flows conspicuously change the dynamics of the debris, while their behaviour has discouraged attempts to use non-invasive techniques such as ultrasonic, X ray, and others (Lee et al. [36]; Abbott et al. [1]).

With regard to the rheological properties, many experiments (Chen [8]) have shown that the Herschel-Bulkley equations fit quite well laboratory data. One of the criticisms that may be moved to these tests is related to the scale effect.

Successful models of debris flows must describe the mechanics of mobilization as well as the subsequent flow and deposition processes. Mobilization requires failure of the mass, a quantity of water to saturate the solid phase, such a change of energy, from gravitational to kinetic, to modify the motion pattern from sliding along a failure surface to a more widespread solid-liquid mixture that can be assessed as flow.

On the whole, laboratory and field data are essential in determining the quality of any mathematical model, as it is especially important for estimating velocity, discharge, concentration, yield stress, viscosity and grain-size (Lorenzini and Mazza [37]). However, the achievement of good agreement between theoretical and experimental results does not justify indiscriminate extrapolation for the various territorial situations, which have very different boundary conditions from standard laboratory conditions. Assuming that the scientific research path cannot exclude an accurate observation and description of the phenomenon in question, without which the analysis of physical processes, that generate it, would become extremely artificial and uncertain, it is hoped that any attempt at improving the interpretation of the phenomenon involves critical comparison between the theoretical, experimental, and field approaches, as well as extensive osmosis process between the same approaches.

4 Debris flows generated by slope failures

Debris flows can be the result of some form of landslides. In particular sliding phenomena in granular soils can turn into flow like movements.

The main difference between slides and flow like landslides concerns the mechanisms of movement. While a slide advances on the slip surface as a rigid block or with a small internal deformation, a flow spreads downslope as a viscous fluid, adapting itself to any morphological change encountered along its path.

In some conditions shear failure (sliding) can be affected by a rapid increase of positive pore pressures in excess to the hydrostatic values. The raise of excess pore pressures decreases the shear resistance of the soil inducing an acceleration of the movement: under these conditions the process can originate a debris flow.

The triggering of positive excess pore pressure in loose granular materials can occur if the soil is saturated and the mechanism of slope deformation is characterized by fast volumetric compression.

The occurrence of flow like movements is dependent on the un-drained behaviour of the soil, which refers to the condition of a saturated soil deforming at constant volume.

The knowledge of the un-drained response of granular soils is of critical importance in assessing their susceptibility to liquefaction. The term liquefaction is frequently used to indicate all phenomena involving excessive deformation in saturated cohesionless soils and is not limited to the development of 100% excess pore pressure. Liquefaction can be triggered by either static or cyclic loading. Liquefaction due to static loading is associated with granular soils deforming in a strain softening (or limited strain softening) manner that results in limited or unlimited unidirectional flow deformation (Sivathayalan and Vaid [48]).

A fundamental understanding of the un-drained response of granular soils has been derived from controlled laboratory studies. Un-drained triaxial compression tests on sand specimens mostly reconstituted by moist tamping have formed the basis for the steady state concepts (Poulos [44]).

Susceptibility of soil to liquefaction mainly depends on grain size and porosity, but also on stress conditions (Picarelli et al. [43]).

Ishihara et al. [30] presented the results of a series of laboratory tests, using triaxial apparatus, on saturated samples of Toyoura sand consolidated anisotropically. They found that with an increasing degree of anisotropy at the time of consolidation the sample becomes more contractive and susceptible to triggering flow failure. They found that the major effective principal stress at the time of anisotropic consolidation is a parameter controlling dilative or contractive behaviour of the sand. As a result the most appropriate way to normalise the residual strength of anisotropically consolidated sand is by the use of major principal stress at consolidation. The quasi steady state strength is then a function of void ratio and the major effective stress at consolidation.

Other contributions devoted to the assessment of the potential for liquefaction of a soil are based on the concept of region of instability (Sladen et al. [49]). Soil instability is a phenomenon that resembles liquefaction in that there is a sudden decrease in the soil strength under un-drained conditions. This loss of strength is related to the development of large pore pressures reducing effective stresses in the soil. Lade [34] showed that there exists a region of instability inside the failure surface. The loss of strength occurs in un-drained condition as a consequence of disturbances small but fast enough to prevent water drainage.

Conventional slope stability analysis methods (limit equilibrium methods) are widely used to investigate landslide problems and to determine the state of stress in slopes. This type of analysis has been used by Lade [34] for the determination of the state of stress in finite slope made of loose sand in order to investigate the region of instability by varying the slope height.

Deangeli [13] presented a study devoted to the assessment of the potential for liquefaction in all zones of finite slopes from the in situ state of stress. For these purposes numerical models reproducing different slopes have been set up by using a finite difference code (FLAC manuals, 2001). The state of stress in slopes has been evaluated in both elastic and elastic-plastic field. By relating this state of stress to the parameters stated by Ishihara et al. [30] to describe the potential for liquefaction a chart of susceptibility of debris flow in soil slopes has been set up. The chart reports curves that establish the limit condition on the basis of critical combinations of void ratio, slope angle and slope height.

In this context some authors define a slope safety factor against liquefaction. For instance Poulos et al. [45] proposed the ratio between the residual strength of the soil S_{us} (the minimum resistance in un-drained conditions for a contractive soil with respect to the in situ void ratio) and the shear stress required for static equilibrium along the potential sliding surface. Ishihara et al. [30] defined the safety factor as the ratio between the residual strength of the soil (which is dependent on the major effective principal stress at the time of anisotropic consolidation) and the maximum shear stress along the potential sliding surface.

Deangeli [13] reported the safety factor against liquefaction along different surfaces passing through a slope and assessed the volume of soil potentially involved in debris flow.

The analysis of the propagation of debris flows generated by slope failures can be performed by taking into account the initial value of excess pore pressure (after slope failure) and its dissipation along the path.

Significant results have been obtained by instrumented laboratory flume experiments. In these experiments the role of pore pressure in the flow failure phase, i.e. the transition from sliding to flow was investigated (Eckersley [20]).

Deangeli [14] set up series of flume experiments to analyze the behaviour of water sand mixture flows, as a consequence of slope failures induced by water table raising and rainfall. The flows initially accelerated but at a certain stage of the process, unsteady deposition of the sand occurred, preceded by the transformation of the movement from flow to sliding. The phenomenon of deposition of the soil along the flume occurred at inclination greater than in the case of Spence and Guymer [51] experiments.

On the basis of the reported results, it is evident the need of further experimental works investigating the dependence of debris flow behaviour by the triggering mechanisms and the role and generation of pore pressure during the propagation phase.

5 Concluding remarks

Debris and hyper-concentrated flow result from the interaction of hydrological processes with geological processes and are triggered when soils get saturated and the stability of the slope is no longer maintained. These flows are among the most destructive of all water-related disasters. In this context, the recognised need to improve knowledge on the mechanics of these solid-liquid flows, highlighted by a critical analysis of the current international state-of-the-art, represent the seeding of the present work.

Although the main aspects that rule the mechanics of these phenomena seem to be understood, it has to be underlined the relative scarcity of experimental (laboratory and field) data, the only ones that allow effective check of the models nowadays available in different flow conditions and the estimation of the rheological parameters they contain.

Greater research needs to be directed towards understanding the nature and the behaviour of these flows. Such knowledge is essential in order to estimate the potential frequency of these natural hazards and design suitable prevention and remediation measures.

The ideal sequence that should be pursued in the approach to the difficult task of the management and mitigation of hyper-concentrated and debris flow can be obtained as follows (De Wrachien [16]):

- first, a systematic collection of field data should be carried out in order to provide a large base of reliable data that could allow a better knowledge of the existing risk trends and a deeper understanding of the mechanics of the phenomena, along with their general behaviour and effects;
- secondly, effective mathematical models, which strongly depend on data and measurements collected and performed in the field for their calibration and design, should be constantly developed, updated when needed, tested and applied;
- hazard mapping techniques and identification of possible scenarios, which need reliable models to be effective and sound, should then be set up;
- on the basis of the knowledge achieved in the previous steps, the best mitigation solutions should be identified, designed and built up;
- finally a program of systematic observations on the sites, where risk has been mitigated, should be planned and carried out to detect any shortcoming and test the efficiency of the investigations.

Each of the above studies and investigations needs improvements and depends, to achieve them, on improvements in other fields. Improving measurement and documentation procedures would provide a better knowledge and ideas for new and more advanced models. The application of existing models based on the data collected in the field and the development of reliable new ones would allow, on one hand, to better focus what to observe in field and, on the other hand, improve mitigation measures and procedures. The field application of these latter would then identify new parameters to be measured and introduced in the models.

From all these activities would emerge the best direction to be followed in future in-depth studies and investigations of debris flows.

References

[1] Abbott J., Mondy L.A., Graham A.L., Brenner H. Techniques for analyzing the behaviour of concentrated suspensions, in *Particulate Two-Phase Flow*, edited by M. C. Roco, pp. 3-32, Butterworth-Heinemann. Newton, Mass., 1993.

[2] Arai M., Takahashi T., The Karman constant of the flow laden with high sediment in *Proc. of the 3rd International Symposium on River Sedimentation* University of Mississippi, 1986, pp. 824-833

[3] Asmar B.N., Lanston P.A., Ergenzinger Z., The potential of the discrete method to simulate debris flow in *Proceeding of the First International Conference on Debris Flow Hazard Mitigation: Mechanics, Prediction and Assessment*, Eds. Chen, New York, 1997

[4] Bagnold R.A., Experiments on a gravity-free dispersion of large solid spheres in a Newtonian fluid under shear in *Proceedings of the Royal Society of London*, Series A, 225, 1954, pp. 49 – 63

[5] Bailard J.A. An experimental study of granular-fluid flow Thesis presented to University of California at San Diego, Calif., 1978

[6] Bellos V., Sakkas J.G., 1D dam – break flood propagation on dry bed *Journal of Hydraulic Engineering*, 1987, ASCE 113(12), pp. 1510 – 1524

[7] Chen C.J., Laboratory verification of a dam – break flood model *Journal of Hydraulic Division ASCE*, 106(4), 1980, pp. 535 – 556

[8] Chen C.L. Bingham plastic or Bagnold dilatant model as a rheological model of debris flow? *Proc. of Third Int. Sympos. on river sedimentation*, University of Mississippi, 31st March – 4th April 1986

[9] Chen L.C., Generalized visco-plastic modelling of debris flow *Journal of Hydraulic Engineering*, 1988, 114, pp. 237 – 258

[10] Chen C.L., Ling C.H., Resistance formulas in hydraulics based models for routing debris flow in *Debris Flow Hazard Mitigation: Mechanics, Prediction and Assessment*, Eds. Chen, New York, 1997, pp. 360 – 372

[11] Courant R., Friedrichs K.O., Supersonic flow and shock wave Interscience Publisher Inc., New York, 1948

[12] Coussot P. Steady, laminar, flow of concentrated mud suspensions in open channel, *Journal of Hydraulic Research*, Vol. 32, n. 4, pp.535-559, 1994

[13] Deangeli C., The Role of Slope Geometry on Flowslide Occurrence, American *Jou. of Environmental Sciences*, Scipub, New York, 3 (3), 2007, pp. 93-97

[14] Deangeli C., Laboratory Granular Flows generated by Slope Failures, *Rock Mechanics Rock Engineering*, Springer, Netherlands, 41 (1) 2008, pp. 199–217

[15] Dent J.D., Lang T.E., A biviscous modified Binghman model of snow avalanche motion *Annals of Glaciology*, 4, 1983, pp. 42 – 46

[16] De Wrachien D. Debris and hyper-concentrated flows, in G. Lorenzini, C.A. Brebbia, D.E. Emmanouloudis (eds) *Monitoring, Simulation, Prevention and Remediation of Dense and Debris Flow*, Rhodes, Greece, 2006

[17] De Wrachien D., Mambretti S. Dam-break shock waves: A two-phase model for mature and immature debris flow *Second International Conference on Debris Flow*, 18 – 20 June 2008, The New Forest, United Kingdom

[18] De Wrachien D., Mambretti S. Dam break with floating debris: a 1D, two-phase model for mature and immature flow propagation *International*

Conference on Agricultural Engineering and Industry Exhibition, 23 – 25 June 2008, Hersonissos, Crete, Greece

[19] Dressler R.F. Hydraulic resistance effect upon the dam-break functions *Proc. of Royal Society of London* A(257), 1952, pp. 185 – 198

[20] Eckersley J.D., Instrumented laboratory flowslides, *Geotechnique*, 40, N. 3, 1990, 489-502.

[21] Fink J.H., Malin M.C., D'Alli R.E., Greeley R. Rheological properties of mudflows associated with the spring 1980 eruptions of Mount St. Helens volcano, *Washington Geophys. Res. Lett.*, 8, 43-46, 1981.

[22] FLAC manuals, 2001, Version 4, ITASCA Consulting group, Minneapolis, USA

[23] Fread D. L., Jin M., One-dimensional Routing of Mud/Debris flows using NWS FLDWAV Model, in *Proc. of First International Conference on Debris Flow Hazards Mitigation: Mechanics, Prediction and Assessment*, San Francisco, California, 7-9 August 1997

[24] Ghilardi P., Natale L., Savi F., Debris flow propagation and deposition on urbanized alluvial fans, *Excerpta*, 14, 2000, pp. 7 – 20

[25] Hanes D.M. Studies on the mechanics of rapidly flowing granular-fluid materials, Thesis presented to Univ. of California at San Diego, Calif., 1983

[26] Harten A. High resolution schemes for hyperbolic conservation laws *Journal of Computational Physics*, 49, 1983, pp. 357-394

[27] Hirano M., Hasada T., Banihabib M.E., Kawahasa K., Estimation of hazard area due to debris flow in *Debris Flow Hazard Mitigation: Mechanics, Prediction and Assessment*, Eds. Chen, New York, 1997, pp. 697-706

[28] Honda N., Egashira S., Prediction of debris flow characteristics in mountain torrents in *Debris Flow Hazard Mitigation: Mechanics, Prediction and Assessment*, Eds. Chen, New York, 1997, pp. 707-716

[29] Hunt B., Asymptotic solution for dam-break problems *Journal of Hydraulic Division ASCE*, 108(1), 1982, pp. 115-126

[30] Ishihara, K., Tsukamoto Y., Shibayama T., Evaluation of slope stability against flow in saturated sand. Reports on Geotechnical engineering, Soil mechanics and Rock engineering, *Jubilee volume of Terzaghi Brandl* 2000. Wien, 2000-2001, Vol. 5, Institut fur Grundbau und Bodenmechanik-Technische Universitat Wien Ed., 2003, pp. 41-54.

[31] Jan C.D., A study on the numerical modelling of debris flow in *Debris Flow Hazard Mitigation: Mechanics, Prediction and Assessment*, Eds. Chen, New York, 1997, pp. 717-726

[32] Johnson A.M. Physical processes in geology Freeman Ed., San Francisco, 1970

[33] Julien P.Y., O'Brien J.S., Physical properties and mechanics of hyperconcentrated sediment flows in Proceeding Spec. *Conference on Delineation of Landslides, Flash Flood and Debris Flow* Utah, USA, 1985, pp. 260-279

[34] Lade P. Static instability and liquefaction of loose fine sandy slopes *J. Geotech. Engng Div ASCE* 118, 1, 1992, 51-71.

[35] Lax P., Wendroff B., Systems of conservation laws *Comp. on Pure and Applied Mathematics* 13, 1960, pp. 217-237
[36] Lee J., Cowin S.C., Templeton III J.S. An experimental study of the kinematics of flow through hoppers. *Trans. Soc. Rit.*, 18, 247-269, 1974.
[37] Lorenzini G., Mazza N. Debris flow. Phenomenology and Rheological Modelling WIT Press, Ashurst Lodge, Southampton, UK, 2004
[38] Mambretti S., Larcan E., De Wrachien D. 1D modelling of dam – break surges with floating debris *Biosystems Engineering*, Vol. 100(2), June 2008, pp. 297-308
[39] McCormack R.W., The effect of viscosity in hypervelocity impact cratering *AIAA Paper*, 1969, 75-1
[40] Norem H., Locat J., Schieldrop B., An approach to the physics and the modelling of the submarine flowslides *Marine Geotechnical* 9, 1990, pp. 93-111
[41] O'Brien J.S., Julien P.J., Fullerton W.T., Two-dimensional water flow and mudflow simulation, *Jou. of Hydraulic Engineering*, 1993, 119, pp. 244-261
[42] Okuda S., Suwa H., Okunishi K., Yokoyama K., Nakano M. Observations on the motion of a debris flow and its geomorphological effects, *J. Geomorphol., suppl.* 35, 142-163, 1980
[43] Picarelli L., Olivares L., Comegna L., Damiano E. Mechanical Aspects of Flow-Like Movements in Granular and Fine Grained Soils, *Rock mechanics rock engineering* Springer, Netherlands, 41 (1) 2008, pp. 179-197.
[44] Poulos S.J., The steady state of deformation. *Jou. of Geotechnical Eng. Div., ASCE*, 107, 1981, pp. 553-561
[45] Poulos S.J., Castro G., France J.W., Liquefaction evaluation procedure, *Jou. Geotechnical. Eng. Div. ASCE*, 111(6), 1985, pp. 772-792.
[46] Ritter A. Die Fortplanzung der Wasserwellen Zeitschrift des Vereines Deutscher Ingenieure 36(3), 1892, pp. 947 – 954 (in German)
[47] Sakkas J.G., Strelkoff T. Dam break flood in a prismatic dry channel *J. Hyd. Div. ASCE* 99(12) 2195-2216, 1973
[48] Sivathayalan, S., Vaid, Y. P. (2002): Influence of generalized initial state and principal stress rotation on the undrained response of sands. *Can. Geotech. Jou.*, 39, 63-76.
[49] Sladen J.A., d'Hollander R.D., Krahm J., The liquefaction of sands, a collapse surface approach, *Can Geotech Jou.*, 22, 1985, pp. 564-578.
[50] Stoker J.J. The breaking of waves in shallow water *Annuals New York Academy of Science* 51(3), 1949, pp.360-375
[51] Spence K.J., Guymer I., Small scale laboratory flowslides, *Geotechnique*, 47, 5, 1997, pp. 915-932.
[52] Strelkoff T., Falvey H.T. Numerical methods used to model unsteady canal flow *J. Irrig. and Drain. Engrg, ASCE*, 119(4), 637-655, 1993
[53] Takahashi T. Debris flow *Rev. Fluid Mechanics*, 13, pp. 57-77, 1981
[54] Takahashi T. Debris Flow, 165 pp., A. A. Balkema, Brookfield. Vt. 1991.
[55] Takahashi T, Initiation of flow of various types of debris flow *Proceeding Second International Conference on Debris Flow Hazard Mitigation:*

Mechanics, Prediction and Assessment, Eds. Wieczorak and Naeser, Rotterdam, 2000, pp. 15-25
[56] Takahashi T., Nakagawa H., Flood / debris flow hydrograph due to collapse of a natural dam by overtopping *Journal of Hydroscience and Hydraulic Engineering*, 1994, 12, pp. 41-49
[57] Van Leer B., Towards the ultimate conservative difference scheme *Journal of Computational Physics* 23, 1977, pp. 263-275
[58] Wan Z. Hyperconcentrated flow Monograph Series of IAHR, Rotterdam, 290 pp., 1994

One-dimensional finite volume simulation of real debris flow events

L. Schippa & S. Pavan
Department of Engineering, Ferrara University, Italy

Abstract

A numerical model for the simulation of mud flow and debris flow is presented. It is based on an alternative formulation of conservative balance equations, in which source terms are mathematically reorganized in order to guarantee an improved computational stability over complex geometry channels. For numerical implementation, the first order Godunov scheme with Roe's approximation is used. Source terms are computed with Euler's method and added by splitting. Such a simple basic scheme has been chosen to underline that the improved numerical stability depends on the proposed mathematical formulation, and not on a sophisticated numerical scheme. The correct wet-dry front velocity and propagation mechanism have been verified with standard dam-break test cases, and particular attention has been directed to the celerity computation inside the Roe's scheme when dealing with irregularly shaped cross-sections. The numerical model has already been verified with analytical tests and laboratory experiments. In this work, the model is applied to two real events that occurred in North-Eastern Italy. The first is a debris flow that took place in the Upper Boite Valley, in the proximity of Cortina d'Ampezzo, in 1998, the second is a mud flow event located in the Stava Creek Valley in 1985. These events have been chosen thanks to the wide documentation and significant amount of field data available, which include topographical surveys, flow velocity measures and flow depth estimations.

Keywords: mud flow, debris flow, wave propagation, source terms.

1 Introduction

The aim of the present work is to check a numerical model that is suitable for the simulation of mud flows and debris flows in channels of complex geometry. To

WIT Transactions on Engineering Sciences, Vol 67, © 2010 WIT Press
www.witpress.com, ISSN 1743-3533 (on-line)
doi:10.2495/DEB100021

fulfill this purpose, the model should have specific features, such as the treatment of wet-dry fronts, the handling of complex geometries and high bed slopes and the possibility of changing the model application field from Newtonian to non-Newtonian fluids, simply by changing the resistance law.

These features have previously been tested applying the model to different test cases that have been properly chosen [1]. The classic frictionless dam-break test has been used to verify the correctness of waves speed propagation and the capability of treating wet-dry fronts. A non-cylindrical frictionless ideal channel has been used to evaluate the model response to abrupt changes in cross-section wideness and bed elevation, then the effect of friction terms introduction has been checked using a mud flow dam-break. The first phase of the model verification ended with the simulation of laboratory experiments on a mud flow dam-break over a sloping plane.

In the present phase, the model is applied to two real events that occurred in North-Eastern Italy. The first is a debris flow that took place in the Upper Boite Valley, in the proximity of Cortina d'Ampezzo, in 1998, the second is a mud flow event located in the Stava Creek Valley in 1985.

The proposed model is based on an alternative formulation of conservative balance equations, which includes a particular mathematical expression of source terms ideated for natural channels, and which has already demonstrated important stability features under the numerical point of view [2, 3]. The numerical implementation is performed using the Godunov finite volumes scheme. This kind of numerical schemes are largely diffused in mud flow or debris flow treatment [4–6], together with the Roe's approximation for the solution of the Riemann problem. The presented model uses the same approach, but paying careful attention in conserving the general formulation suitable for complex geometry channels, in particular for what concerns the expression of the wave propagation celerity. This term is usually expressed as a function of water depth and cross-section width, but these hydraulic quantities often need to be corrected or mediated to be representative of irregular cross-sections. As an alternative, cross-section shape can be parameterized to be numerically handled [7]. In this work, celerity is determined referring to cross section wetted area and static moment, in order to ensure the formulation generality.

Source terms are handled using the splitting technique [8] and evaluated with the Euler's method. The pressure source terms, induced by the channel irregular geometry have been treated as in [2, 3], mathematically transforming the derivative of the static moment in order to eliminate the explicit dependence on the channel bed slope. This operation keeps its validity also in case of highly sloping channels, condition which often occurs in mud flow or debris flow phenomena. Friction source terms depends on the evaluation of friction slope, and therefore on the adopted resistance law. Like most of numerical models [5], the proposed model set up permits to easily change the resistance law and therefore to use the best fitting rheological model for each test case. It is worth noting that source terms numerical implementation has been kept as simple as possible, to put in evidence the stability features coming from the basis mathematical model.

2 Mathematical model

The mathematical model is based on an alternative formulation of shallow water equations for one-dimensional (1-D) flows in natural channels of complex geometry [2]. The continuity equation and the momentum balance equation are written in terms of state variables A and Q, considering no lateral inflows.

$$\frac{\partial A}{\partial t}+\frac{\partial Q}{\partial x}=0 \tag{1}$$

$$\frac{\partial Q}{\partial t}+\frac{\partial}{\partial x}\left(\frac{Q^2}{A}+gI_1\right)=g\left.\frac{\partial I_1}{\partial x}\right|_{z_w}-gAS_f \tag{2}$$

where t is time, x is distance along the channel, A the wetted cross-sectional area, Q the discharge, g the gravitational acceleration, I_1 the static moment of the wetted area, defined as:

$$I_1=\cos\vartheta\int_0^{h(x)}b(x,z)\left(h(x)-z\right)\mathrm{d}z \tag{3}$$

I_2 is the variation of the static moment I_1 along the x-direction, $S_o=\sin\vartheta$, where ϑ is the angle between channel bottom and the horizontal, b is the cross-section width, h is flow depth.

The system closure equation for the evaluation of the friction term S_f will be described in detail for each examined test case, but the generally considered formulation is

$$S_f=\frac{\tau}{\rho gR} \tag{4}$$

in which S_f is the slope friction, R is the hydraulic radius, ρ is the mixture or the fluid density, and the shear stress τ depends on the adopted rheological model.

2.1 The source term

Differently from the commonly used formulation of shallow water equations, the proposed model does not include in the momentum balance equation source term a direct dependence on bed slope. Details on the mathematical treatment which led to eqn. (2) can be found in [3].

The classic momentum equation is

$$\frac{\partial Q}{\partial t}+\frac{\partial}{\partial x}\left(\frac{Q^2}{A}+gI_1\right)=gA\left(S_0-S_f\right)+gI_2 \tag{5}$$

Focusing on the source term, the pressure term I_2 has the following expression:

$$I_2=\left.\frac{\partial I_1}{\partial x}\right|_h=\cos\vartheta\int_0^{h(x)}\left(h(x)-z\right)\frac{\partial b(x,z)}{\partial x}\mathrm{d}z \tag{6}$$

Briefly, the pressure term I_2 can be expressed as the sum of two terms, one of which is the variation of static moment I_1 along x considering the water surface elevation z_w as a constant, while the other exactly balances gravitational forces in the momentum equation, unless the presence of the term $\cos\vartheta$ which arises in case of high slopes, and cannot be neglected when considering mud-flow or debris-flow phenomena.

$$I_2 = \left.\frac{\partial I_1}{\partial x}\right|_{z_w} - S_0 A \cos \vartheta \tag{7}$$

The substitution into (6) produces:

$$gA\left(S_0 - S_f\right) + gI_2 = gAS_0\left(1-\cos\vartheta\right) - gAS_f + g\left.\frac{\partial I_1}{\partial x}\right|_{z_w} \tag{8}$$

In this case, the term AS_0 does not disappear as illustrated in [2, 3], but it remains and it is multiplied by the factor (1-cosϑ). However, numerical proofs have demonstrated that this term is little if compared to friction terms, and can therefore be neglected. Eqn. (2) is therefore valid also for high sloping channel and debris flow simulation.

3 Numerical model

Shallow water equations have been numerically implemented using the first-order finite volumes Godunov scheme. Numerical fluxes are computed with Roe's method and source terms are evaluated with Euler's approach and taken into account adopting the splitting technique. Details on the different components of the numerical model can be found in Toro [8]. The resultant scheme is explicit, first-order accurate, and has a very uncomplicated structure, since it is built choosing the simplest solution technique for every element of the partial differential equations system. This approach has the intention to illustrate the intrinsic stability features of the mathematical model, which could otherwise be hidden by sophisticated numerical schemes.

Referring to shallow water equations in the vector form (eqn. (9)) the splitting approach for source terms treating, consists in separately solving the homogeneous partial differential equations system (eqn. (10)) and the ordinary differential equation (eqn. (11)). In detail, the solution obtained from eqn. (10) is used as initial condition for eqn. (11).

$$\mathbf{U}_t + \mathbf{F}\left(\mathbf{U}\right)_x = \mathbf{S}\left(\mathbf{U}\right) \tag{9}$$

$$\mathbf{U}_t + \mathbf{F}\left(\mathbf{U}\right)_x = 0 \;\Rightarrow\; \bar{\mathbf{U}} \tag{10}$$

$$\mathbf{U}_t = \mathbf{S}\left(\bar{\mathbf{U}}\right) \;\Rightarrow\; \mathbf{U}_{t+dt} \tag{11}$$

The Roe's scheme, used to solve eqn. (6), requires the definition of the Jacobian matrix

$$\mathbf{J} = \frac{\partial \mathbf{F}}{\partial \mathbf{U}} = \begin{pmatrix} 0 & 1 \\ g\dfrac{\partial I_1}{\partial A} - \dfrac{Q^2}{A^2} & 2\dfrac{Q}{A} \end{pmatrix} = \begin{pmatrix} 0 & 1 \\ c^2 - u^2 & 2u \end{pmatrix} \tag{12}$$

Most of models proposed in the literature about the resolution of shallow water equations for debris flow or natural channels, based on approximate Riemann solvers (see for example [4, 5, 9]), adopt the same simplification in the evaluation of the term $\partial I_1/\partial A$, assuming

$$\frac{\partial I_1}{\partial A} \simeq \frac{A}{B} \simeq h \quad \Rightarrow \quad c = \sqrt{g\frac{A}{b}} \quad \text{or} \quad c = \sqrt{gh} \tag{13}$$

In the present model, in order to keep the formulation generality and to ensure the applicability to natural and complex channel geometries, the static moment derivative is explicitly computed as the variation of I_1 relative to the variation of A in the water depth variation range $h \pm \Delta h$

$$\frac{\partial I_1}{\partial A} = \frac{I_1(h+\Delta h) - I_1(h-\Delta h)}{A(h+\Delta h) - A(h-\Delta h)} \tag{14}$$

The celerity c is therefore defined as

$$c = \sqrt{g\frac{\partial I_1}{\partial A}} \tag{15}$$

Another important aspect of the Godunov finite volume method application to natural geometries is the quantification of cell water volume V and the definition of the relation between the state variable A and V. For every computational cell, A is defined as

$$A_i = \frac{1}{\Delta x}\int_{x_{i-\frac{1}{2}}}^{x_{i+\frac{1}{2}}} A(x,t)\,\mathrm{d}x = \frac{V_i}{\Delta x} \tag{16}$$

V_i is computed as the volume of a pyramid which bases are irregular polygons, since the water profile is assumed to be parallel to channel bed.

$$V_i = \frac{\left(A_{i-\frac{1}{2}} + A_{i+\frac{1}{2}} + \sqrt{A_{i-\frac{1}{2}}A_{i+\frac{1}{2}}}\right)\Delta x}{3} \tag{17}$$

3.1 Source terms numerical treatment

Source terms are numerically included in computations by splitting, and they are simply computed by Euler's method

$$\mathbf{U}_{t+dt} = \mathbf{U}_t + \Delta t \cdot \mathbf{S}(t, \bar{\mathbf{U}}) \tag{18}$$

in which

$$\mathbf{S} = \begin{pmatrix} 0 \\ g\left.\frac{\partial I_1}{\partial x}\right|_{z_w} - gAS_f \end{pmatrix} \tag{19}$$

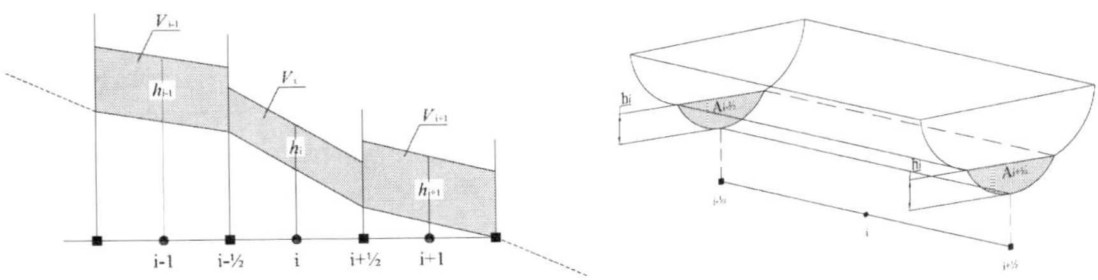

Figure 1: Computational scheme for V_i.

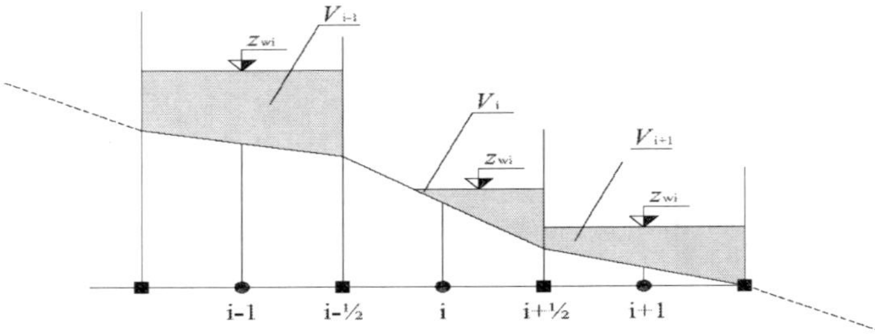

Figure 2: Computational scheme for momentum balance pressure source term.

Considering no lateral inflows, source terms are present only in the momentum balance equation. This term can be divided into two parts, that is the friction term and the pressure term, represented by the static moment variation along channel, taking the water surface elevation as a constant.

The computational scheme for the pressure term quantification in represented in Figure 2, and the variation of I_1 is computed as:

$$\left.\frac{\partial I_1}{\partial x}\right|_{z_w} \simeq \left.\frac{\Delta I_1}{\Delta x}\right|_{z_w} = \frac{I_1\left(h_{i+\frac{1}{2}}\right) - I_1\left(h_{i-\frac{1}{2}}\right)}{\Delta x} \tag{20}$$

4 Numerical tests

In this work the model has been applied to two real events. The first is a natural debris flow event, due to intense rainfall, surveyed at the Acquabona site in Northern Italy. It is of particular interest thanks to the large amount of available field data. The second is the Stava mud flow, a tragic episode occurred in a little town of Italian Alps. This event was caused by the collapse of two tailing dams, which released a huge quantity of water into the Stava Creek channel, causing the formation of a mud flow wave with an enormous destructive power.

4.1 Acquabona debris flow

The Acquabona debris flow has been widely surveyed and documented in the context of the “Debris Flow Risk” Project, funded by the EU. In particular, the UPD (resp. Prof. Rinaldo Genevois) has carried out a research on some debris flow prone watersheds in the Upper Boite Valley (Eastern Dolomites, Southern Alps) and surroundings, included in the municipality of Cortina d’Ampezzo [10]. A large quantity of field data is therefore available since an automatic, remotely controlled monitoring system has been installed at Acquabona on June 1997. The Acquabona site in characterized by one or more debris flow every year, which usually occur in summer and in early autumn and are associated to intense, spatially limited rainfall events. The monitoring system installed at Acquabona was fully automatic and remotely controlled. It consisted of three on-site monitoring stations and an off-site master collection station. Every station was equipped with a geophone, while at Station 3 also a superficial pressure transducer and an ultrasonic sensor were present.

In this work we refer to the event of the August 17th in 1998. The event was originated by a very intense rainstorm: 25.4 mm of rain were measured during 30 min by the rain-gage at Station 1. The volume of the deposits available for debris flow generation has been estimated to be around 8000-9000 m^3. The overall duration of the event was of approximately 38 min and more than 20 different surges have been surveyed at Station 3.

The geometry of the channel is available thanks to 19 surveyed transversal cross-sections, for a global channel length of 1120 m and a difference in height of 245 m. The longitudinal slope ranges from 10% to 55%. For model application a constant spatial step of 1 m has been adopted. Numerical simulations were performed adopting the rainfall hydrograph reconstructed by Orlandini and Lamberti [11], which has an extension of about 2.5 hours and a peak discharge of 2.3 m^3/s. An open boundary type condition is imposed at the downstream end. For the debris flow the bulk concentration is assumed to be 0.6 and mixture density 1850 kg/m^3, according to [7].

The rheological model adopted in the simulations is the Herschel-Bulkley model, which, for simple shear conditions may be written as:

$$\tau = \tau_c + K\gamma^\eta \qquad (21)$$

in which K and η are rheological parameters. Referring to the simulations carried out by Fraccarollo and Papa [12] on the same event, K is assumed to be 150 Pa·s$^{1/3}$, τ_c is equal to 925 N/m^2, and η has been empirically set equal to 1/3.

In Figure 3 computed flow height is compared with the measured data collected by the ultrasonic sensor at Station 3. The model satisfactorily captures wave height and shape, but it underestimates their duration, overestimating as a consequence their number. Results arte however encouraging and comparable to those obtained by Fraccarollo and Papa [12] and Zanuttigh and Lamberti [7].

The average velocity of the different flow surges has been estimated through geophone log recordings. Available data refer to two 100 m channel reaches

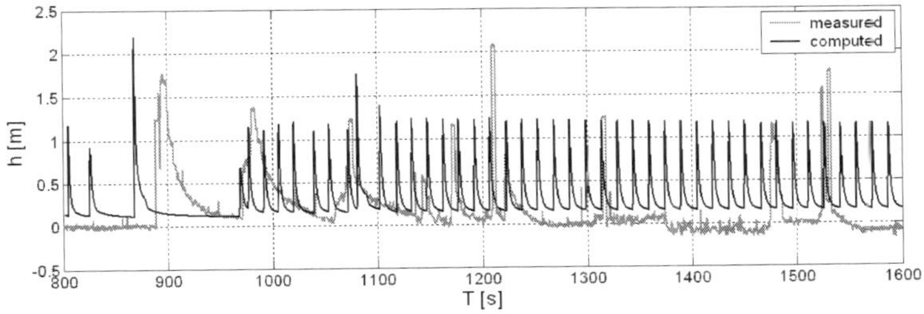

Figure 3: Comparison between the flow depth measured and calculated at Station 3.

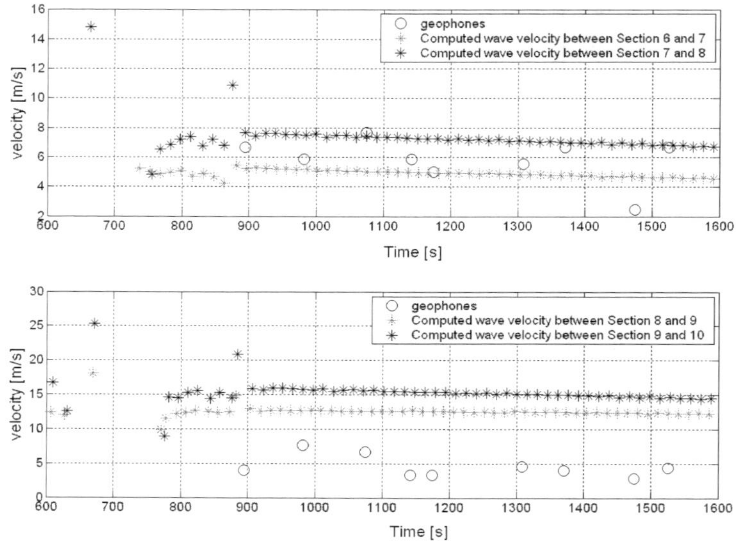

Figure 4: Comparison between measured and computed wave speeds upstream and downstream from Station 3.

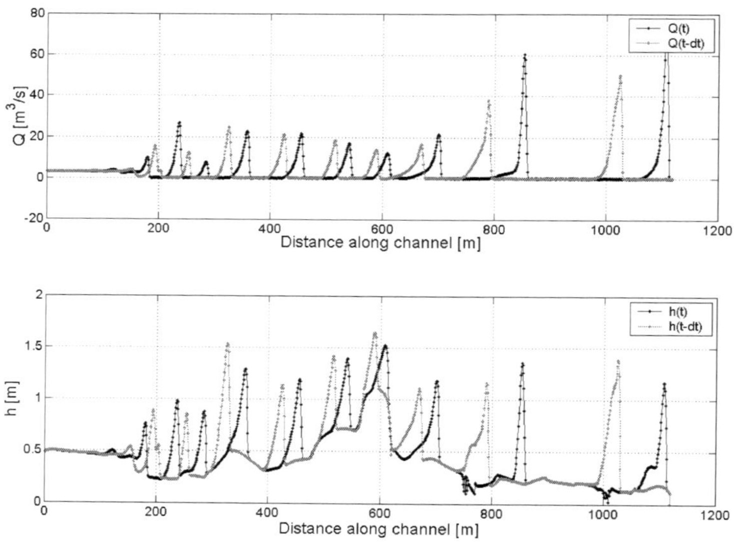

Figure 5: Longitudinal discharge distribution and flow depth profile.

located in the lower part of the channel before and after Station 3, which corresponds to the surveyed cross-section 8. Comparison is showed in Figure 4. In the upstream reach computed velocities compare well with field data, while in the downstream reach they are generally overestimated.

It is interesting noting that the flow regime is mainly characterized by the formation of roll waves, as it is evident observing the longitudinal distribution of discharges and flow depths at two subsequent time steps. Nevertheless, numerical solution is not affected by relevant numerical instabilities.

4.2 Stava mud flow

In July 19th 1985, two tailing dams suddenly collapsed in Tesero, a little town in the Italian Alps. The stored water, together with the dam body material flowed down to the Stava River as a big mud flow, claiming 268 human lives and destroying 47 houses. As reported by Takahashi [13], the Stava River before the disaster flowed with an approximately uniform slope of 5°. Although the mud flow had such an intensive destructive power, as well as fluidity, the Stava River channel itself had not suffered much erosion or deposition, and it can therefore be simulated as a fixed bed stream. In his report Takahashi gives important references also about mud flow solids concentration which was as high as 0.5, while the particle size was so fine that the relative depth, R/d, had a value of the order of 10^5. In this condition the resistance to flow is similar to that of a plain water flow and the Manning's equation can be applied. Takahashi obtained a Manning's roughness coefficient in each section by reverse calculation from the data on velocity computed with the Lenau's formula applied to measured flow superelevations at bends.

The channel description is also taken from Takahashi [13]. It includes 24 surveyed cross-sections, their planimetric location and the longitudinal profile. In this case bed slope ranges from 5% to 12%. The simulated reach is 3500 m long and a constant spatial step of 1.25 m has been used.

In Figure 6, discharge and depth computed hydrographs are compared with Takahashi numerical results obtained with the kinematic wave theory [13]. Referring to cross-section 10, located about 3000 m downstream the dams, there is a good accordance between the computed peak discharge and the value estimated by Takahashi (3500 m^3/s) as a result of product between the wetted cross-section area measured in situ (about 500 m^2) and the maximum velocity derived by the flow superelevation at the nearest bend (7 m/s)

The initial water profile condition reproduces the same hypothesis adopted by Takahashi, which is a uniform slide of the mud mass until Section 4, from which the mud flow is assumed to develop.

Figure 8 shows the comparison between computed and measured front arrival times at different locations. The measured values are estimated on the basis of a seismograph located in Cavalese, a nearby town. The computed times are in good agreement with the estimated ones along the entire channel.

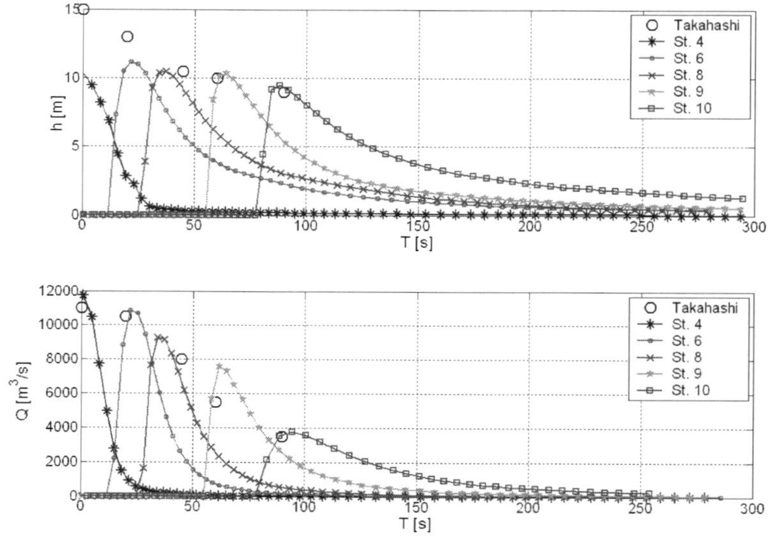

Figure 6: Depth and discharge hydrograph at different cross sections.

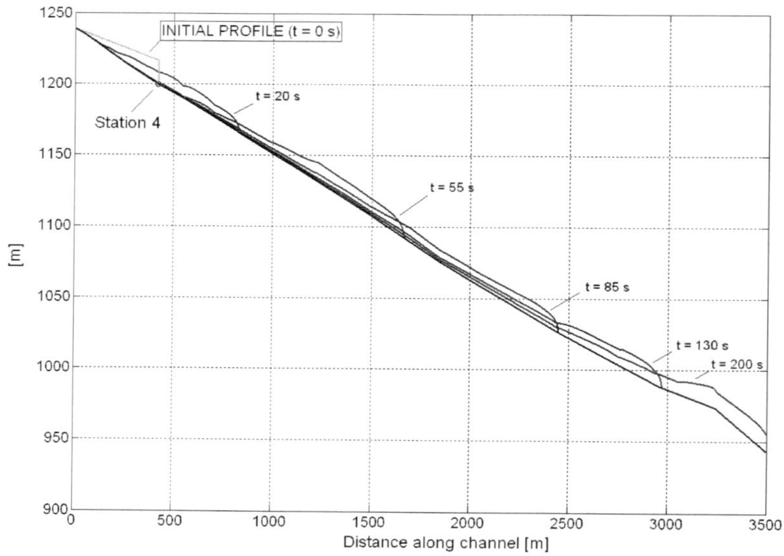

Figure 7: Initial conditions and flow profiles along the channel during the simulation.

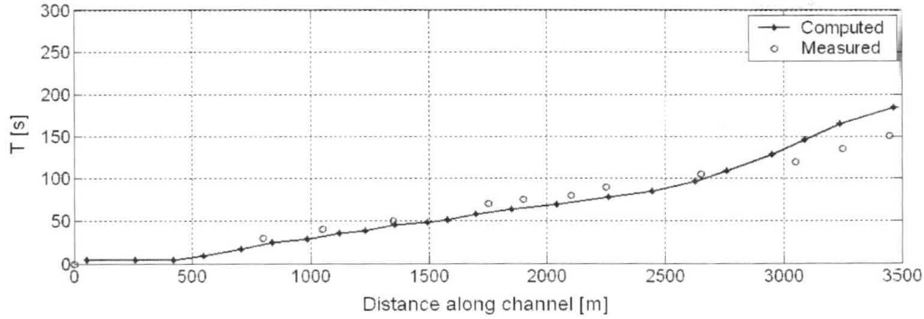

Figure 8: Comparison between computed and measured front arrival times at different locations.

5 Conclusions

A numerical model for the simulation of mud flow and debris flow natural events is presented. It is based on a mathematical model which main features are concerned with the propagation of the wet-dry fronts, the treatment of irregular and variable cross sections shape, and the applicability to highly sloping channels. Two real events have been chosen to test the model. The first is a natural debris flow event at Acquabona site. In this case a large quantity of field data was available and model results compared well with wave peak height and propagation velocities. The second test case refers to the Stava mud flow tragic event, originated by the collapse of two tailing dams. Also in this case good accordance between observed data and mud front propagation speed has been obtained. Simulation results have also been compared with the Takahashi analysis of the same event, showing good accordance for what concerns peak discharge estimation at different cross sections.

References

[1] Schippa., L. & Pavan, S. 1-D finite volume model for dam-break induced mud-flow. *River Basin Management V*, 07-09 September 2009, Malta, pp. 125-136, ed. C.A. Brebbia, Wit Press, Southampton, Boston, 2009.

[2] Schippa., L. & Pavan, S., Analytical treatment of source terms for complex channel geometry. *Journal of Hydraulic Research*, **46(6)**, pp. 753-763, 2008.

[3] Schippa., L. & Pavan, S., Bed evolution numerical model for rapidly varying flow in natural streams. *Computer & Geosciences*, **35**, pp. 390-402, 2009.

[4] Garcia-Navarro, P. & Vazquez-Cendon M.E., On numerical treatment of the source terms in the shallow water equations. *Computer & Fluids*, **29**, pp. 951-979, 2000.

[5] Brufau, P., Garcia-Navarro, P., Ghilardi, P., Natale, L. & Savi, F., 1D Mathematical modelling of debris flow. *Journal of Hydraulic Research*, **38(6)**, pp. 435-446, 2000.
[6] Naef, D., Rickenmann, D., Rutschmann, P. & McArdell, B.W., Comparison of flow resistance relations for debris flow using a one-dimensional finite element simulation model., *Natural Hazards and Earth System Sciences*, **6**, pp.155-165, 2006.
[7] Zanuttigh, B. & Lamberti, A., Analysis of debris wave development with one-dimensional shallow-water equations, *Journal of Hydraulic Engineering*, **130(4)**, pp. 293-303, 2004.
[8] Toro, E.F., *Riemann Solvers and Numerical Method for Fluid Dynamics*, Springer-Verlag Berlin Heidelberg New York, 1999.
[9] Ying, X. & Wang, S.S.Y., Improved implementation of the HLL approximate Riemann solver for one-dimensional open channel flows. *Journal of Hydraulic Research*, **46(1)**, pp. 21-34, 2008.
[10] Berti, M., Geneovis, R., Simoni, A. & Tecca, P.R., Field observations of a debris flow event in the Dolomites., *Geomorphology*, **29**, pp. 265-274, 1999.
[11] Orlandini, S. & Lamberti A., Effect of wind precipitation intercepted by steep mountain slopes. *Journal of the hydrologic engineering*, **5(4)**, pp. 346-354, 2000
[12] Fraccarollo, L., & Papa, M., Numerical simulation of real debris-flow events. *Physics and Chemistry of the Earth*, **25(9)**, pp. 757-763, 2000.
[13] Takahashi T., Debris flow, *IAHR Monograph Series*, A.A. Balkema Rotterdam Brookfield, 165 pp, 1991.

Debris flow modelling accounting for large boulder transport

C. Martinez[1], F. Miralles-Wilhelm[1] & R. Garcia-Martinez[2]
[1]*Department of Civil and Environmental Engineering, Florida International University, USA*
[2]*Applied Research Center, Florida International University and FLO-2D Software, Inc., USA*

Abstract

We present a quasi three-dimensional numerical model to simulate stony debris flows, considering a continuum fluid phase of water and fine sediments, and a non-continuum phase of large particles, such as boulders. Large particles are treated in a Lagrangian frame of reference using the Discrete Element Method in three dimensions. The fluid phase is governed by the depth-averaged Navier–Stokes equations in two horizontal dimensions and is solved by the Finite Element Method. The model simulates particle-particle collisions and wall-particle collisions, taking into account that particles are immersed in the fluid. Bingham and Cross rheological models are used for the continuum phase. Both formulations provide stable results, even in the range of very low shear rates. The Bingham formulation is better able to simulate the stopping stage of the fluid. The results of the numerical simulations are compared with data from laboratory experiments on a flume-fan model. The results show that the model is capable of simulating the motion of big particles moving in the fluid flow, handling dense particulate flows that avoid overlapping among particles. An application to simulate a debris flow event that occurred in Northern Venezuela in 1999 shows that the model replicates well the main observed boulder accumulation areas.

Keywords: debris flow, mud flow, boulders transport, Eulerian and Lagrangian formulation, finite element method, discrete element method.

WIT Transactions on Engineering Sciences, Vol 67, © 2010 WIT Press
www.witpress.com, ISSN 1743-3533 (on-line)
doi:10.2495/DEB100031

1 Introduction

Debris flow is a frequent phenomenon in mountainous regions. It occurs when masses of poorly sorted sediments, rocks and fine material, agitated and mixed with water, surge down slopes in response to water flow and gravitational attraction. A typical surge of debris flow has a steep front or "head" with the densest slurry, the highest concentration of boulders and the greatest depth. A progressively more dilute and shallower tail follows this head.

Reviews presented by Iverson [1], exhaustively describe the physical aspects of debris flow motion and clearly divide previous debris flow research into two distinct categories. The first, based upon the pioneering work of Johnson [2], assumes that debris flow behaves as a viscoplastic continuum. This model describes a single-phase material that remains rigid unless stresses exceed a threshold value: the plastic yield stress. Various rheological models have been proposed, derived from experimental results or from theoretical considerations, such as the Bingham model [3], the Cross model [4], and the quadratic model proposed by O'Brien and Julien [5]. The Bingham plastic model is the most commonly used in practice.

The second approach has focus on the mechanics of granular materials. Based upon the findings of Bagnold [6], two-phase models have been developed by several authors, such as Takahashi [7] and Iverson [1]. These models explicitly account for solid and fluid volume fractions and mass changes respectively.

Despite of the considerable progress over the past few years, the flow dynamics and internal processes of debris flows are still challenging in many aspects. In particular, there are many factors related to the movement and interaction of individual boulders and coarse sediments that have not been fully addressed in previous works. Asmar *et al.* [8] introduced the Discrete Element Method (DEM) to simulate the motion of solid particles in debris flows. DEM is a numerical method to model dry granular flows where each particle is traced individually in a Lagrangian frame of reference by solving Newton's equation of motion.

This paper describes the development of a quasi three-dimensional model to simulate stony debris flows, considering a continuum fluid phase, and large sediment particles, such as boulders, as a non-continuum phase. Large particles are treated in a Lagrangian frame of reference using DEM, and the fluid phase composed by water and fine sediments is modelled with an Eulerian approach using the depth-averaged Navier–Stokes equations in two dimensions. Bingham and Cross rheological models are used for the continuum phase. Particle's equations of motion are fully three-dimensional. The model is tested with laboratory experiments and with a real application.

2 Governing equations

The flow domain is divided in computational cells with triangular base and depth H, as shown in Figure 1.

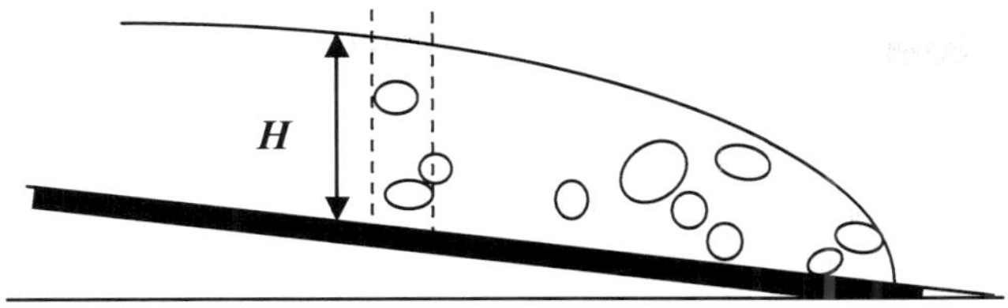

Figure 1: Schematic representation of debris flow with large solid particles.

Assuming non-Newtonian and incompressible fluid phase, the depth averaged continuity and momentum equations in Cartesian coordinates can be written as follows:

$$\frac{\partial H}{\partial t}+\frac{\partial(\overline{u}H)}{\partial x}+\frac{\partial(\overline{v}H)}{\partial y}=0 \tag{1}$$

$$\frac{1}{g}\frac{\partial\overline{u}}{\partial t}+\frac{\overline{u}}{g}\frac{\partial\overline{u}}{\partial x}+\frac{\overline{v}}{g}\frac{\partial\overline{u}}{\partial y}+\frac{\partial\eta}{\partial x}+\frac{F_{Dx}}{\rho g}+S_{fx}=0 \tag{2}$$

$$\frac{1}{g}\frac{\partial\overline{v}}{\partial t}+\frac{\overline{u}}{g}\frac{\partial\overline{v}}{\partial x}+\frac{\overline{v}}{g}\frac{\partial\overline{v}}{\partial y}+\frac{\partial\eta}{\partial y}+\frac{F_{Dy}}{\rho g}+S_{fy}=0 \tag{3}$$

where H is the water depth, η is the free-surface elevation, $\overline{u}$ and $\overline{v}$ are the depth averaged velocities in x and y directions respectively, g is the gravitational acceleration and ρ is fluid density. $\mathbf{F}_D$ represents the fluid-solid interaction force exerted on the fluid by particles through the fluid drag force.), this force is evaluated as:

$$\mathbf{F}_D=\frac{\sum_{i=1}^{n}\mathbf{F}_{FDi}}{\Delta V} \tag{4}$$

where $\mathbf{F}_{FD}$ is the fluid drag force on each particle i, ΔV is the volume of the computational cell and n is the number of particles in the cell. S_{fx} and S_{fy} are the depth integrated stress terms that depend on the rheological formulation used to model the slurry.

Assuming a Bingham rheological model and Manning's formula, as proposed by O'Brien and Julien [5], the stress terms for the fluid can be expressed as

$$S_{fx} = \frac{\tau_y}{\rho g H} + \frac{3\mu\bar{u}}{\rho g H^2} + \frac{N^2 \bar{u}^2}{H^{4/3}} \tag{5}$$

$$S_{fy} = \frac{\tau_y}{\rho g H} + \frac{3\mu\bar{v}}{\rho g H^2} + \frac{N^2 \bar{v}^2}{H^{4/3}} \tag{6}$$

where N is the Manning roughness coefficient.

The fluid dynamic viscosity μ and yield stress τ_y, are determined as functions of the volume sediment concentration Cv, using the relationships proposed by O'Brien and Julien [9]:

$$\mu = \alpha_1 e^{\beta_1 \bar{c}} \tag{7}$$

$$\tau_y = \alpha_2 e^{\beta_2 \bar{c}} \tag{8}$$

in which α_1, β_1, α_2 and β_2 are empirical coefficients obtained by data correlation in a number of experiments with various sediment mixtures. Using a quadratic formulation combined with the Cross rheological model, the stress terms for the fluid are expressed as

$$S_{fx} = \frac{\mu_{eff}\dot{\gamma}}{\rho g H} + \frac{N^2 \bar{u}^2}{H^{4/3}} \quad \text{with } \dot{\gamma} = \frac{3\bar{u}}{H} \tag{9}$$

$$S_{fy} = \frac{\mu_{eff}\dot{\gamma}}{\rho g H} + \frac{N^2 \bar{v}^2}{H^{4/3}} \quad \text{with } \dot{\gamma} = \frac{3\bar{v}}{H} \tag{10}$$

where μ_{eff} is the effective viscosity of the fluid defined by:

$$\mu_{eff} = \frac{\mu_0 + \mu_\infty K_B \dot{\gamma}}{1 + K_B \dot{\gamma}} \tag{11}$$

with $K_B = \dfrac{\mu_0}{\tau_y}$, $\mu_\infty = \mu$ and $\mu_0 = 10^3 \mu$

In the solid phase, spherical particles of different diameters are considered. Particle trajectories are tracked using Newton's second law and the considering gravity, buoyancy, fluid drag and collision forces.

$$m_i \frac{d\mathbf{v}}{dt} = \sum \mathbf{F}_E + \sum \mathbf{F}_N + \sum \mathbf{F}_T \tag{12}$$

The external force $\mathbf{F}_E$ is given by

$$\mathbf{F}_E = \mathbf{F}_B + \mathbf{F}_{FD} \tag{13}$$

The expression to compute the net force acting on the particle due to gravitational effects is

$$\mathbf{F}_B = \frac{4}{3}\pi R^3 (\rho - \rho_p)\mathbf{g} \tag{14}$$

where R is the particle radius and ρ_p is the particle density.

The expression for the drag on particles in viscous fluid is given by

$$\mathbf{F}_{FD} = \frac{1}{2}\pi R^2 C_d \rho |\mathbf{u} - \mathbf{v}| (\mathbf{u} - \mathbf{v}) \tag{15}$$

where C_d is the drag coefficient, **u** is the fluid velocity vector at the particle location, and **v** is the particle velocity vector.

The last two terms in equation (12) represent the collision forces or contact forces among particles. Based on the simplified model that uses a spring-dashpot-slider system to represent particle interactions [8], the normal contact force and the tangential contact force are evaluated as

$$\mathbf{F}_N = \mathbf{F}_{NC} + \mathbf{F}_{ND} \tag{16}$$

$$\mathbf{F}_T = \mathbf{F}_{TC} + \mathbf{F}_{TD} \tag{17}$$

The normal contact force F_{NC} is calculated using a Hook's linear spring relationship,

$$\mathbf{F}_{NC} = K_N \delta_N \tag{18}$$

where K_N is the normal contact stiffness and δ_N is the displacement (overlap) between particles i and j.

The normal damping force F_{ND} is also calculated using a linear relation given by

$$\mathbf{F}_{ND} = C_N v_N \tag{19}$$

where v_N is the normal component of the relative velocity between particles and C_N is the normal damping coefficient. This constant C_N is chosen to give a required coefficient of restitution β, defined as the ratio of the normal component of the relative velocities before and after collision.

The tangential contact force, FTC, represents the friction force and it is constrained by the Coulomb frictional limit, at which point the particles begin to slide over each other. Prior to sliding, the tangential contact force is calculated using a linear spring relationship,

$$\mathbf{F}_{TC} = K_T \delta_T \tag{20}$$

where K_T is the tangential stiffness coefficient, and δ_T is the total tangential displacement between the surfaces of particles i and j since their initial contact. When $K_T\delta_T$ exceeds the frictional limit force $\mu_f F_{NC}$, particle sliding occurs. The sliding condition is defined as

$$\mathbf{F}_{TC} = \mu_f \mathbf{F}_{NC} \tag{21}$$

where μ_f is the dynamic friction coefficient.

The tangential damping force $\mathbf{F}_{TD}$ is not included in this model, since it is assumed that once sliding occurs, damping is accounted for from friction.

Also, particle rotation is not considered. Fluid governing equations (1-3) are solved by the Galerkin Finite Element method using three-node triangular elements. To solve the resulting system of ordinary differential equation, the model applies a four-step time stepping scheme and a selective lumping method, as described by Garcia-Martinez *et al.* [10].

Forces on each particle are evaluated at each time step, and the acceleration of the particle is computed from the particle governing equation, which is then integrated to find velocity and displacement of each particle.

3 Results

A series of experiments were carried out in a laboratory flume, using homogeneous fluid and fine sediment mixtures for the continuum phase and spherical marbles for the discrete phase. The experiments were performed in a 1.9 m long, 0.19 m wide, Plexiglas walled flume, with adjustable slope. The downstream part of the flume was connected to a wood horizontal platform, 0.75 m long and 0.95 m wide. A dam-break type of flow was initiated by an abrupt removal of a gate releasing mixtures from a 0.40 m long reservoir situated on the upstream part of the flume. Water-clay mixtures were used in all the experiments, with volume sediment concentration 23.5% and 26.5%. For preparation of the mixtures, kaolinite clay with specific unit weight of 2.77 was used. Fluid density was measured in the laboratory and rheological parameters μ and τ_y were determined using equations (7) and (8) in which parameters are $\alpha_1 = 0.621\text{x}10^{-3}$, $\beta_1 = 17.3$, $\alpha_2 = 0.002$ and $\beta_2 = 40.2$.

Table 1: Rheological properties of experimental fluids.

Cv (%)	ρ (Kg/m^3)	μ (Pa.s)	τ_y (Pa)
23.5	1410	0.0362	25.34
26.5	1460	0.0608	84.64

3.1 Experiment 1

In this experiment, the flow of a fluid of 23.5% volume concentration was studied. The flume bottom slope was set to 9.54° and the initial volume released was 6.4 L. The objective of this test was to study the spreading of the fluid in the fan and the particle interaction with the fluid.

14 particles, with diameter D = 2.5 cm and density ρ_p = 2500 Kg/m^3, were placed over a small piece of wood inside the mud reservoir, just behind the gate. By the time the fluid was released, the piece of wood was quickly removed, so that the particles could start their movement along the flume with the fluid. At early times after the release, particles travelled downstream on a parabolic formation across the flume following the parabolic velocity profile. However, as the flow moved downstream, particles tended to move to the flume sides.

It is important to mention the effect of the boundary condition for velocity on the sidewalls. Typically, this condition should be a no-slip condition expressed as $\vec{u} = 0$ at the wall. However, in reality, the velocity gradient near the wall is large and the velocity near a wall quickly becomes non-zero. Therefore, in practical applications, this condition becomes very restrictive, causing unrealistic delay of the flow. For this simulation, a more relaxed boundary condition was tested where, the normal velocity is $u_N = 0$, and the tangential velocity is $u_T = 0.9(\vec{u} \cdot \vec{t})$.

Figure 2 compares final position of particles obtained numerically, with observed final position for particles (t = 10 s). The flood extent and the final particle locations obtained numerically replicates reasonably well the experiment. In the numerical results there is some delay on the particles positioned close to the walls that is attributed to the calibration of the boundary condition for tangential velocity at the walls. If a full slip condition is imposed, then the transversal velocity profile disappears across the channel, generating unrealistic results as shown in Figure 3.

3.2 Experiment 2

In this experiment, a mixture of volumetric concentration of 26.5% was studied. In this case, the flume bottom slope was increased to 10.7° and the initial volume released was 11.1 L. The objective of this test was to study the spreading of the fluid and study particle movement into a mixture with higher clay concentration.

In this experiment, the velocity of the front wave is basically constant until reaching 1.6 m, from this point the celerity of the wave decreases abruptly, taking about 40 s for the fluid to stop completely.

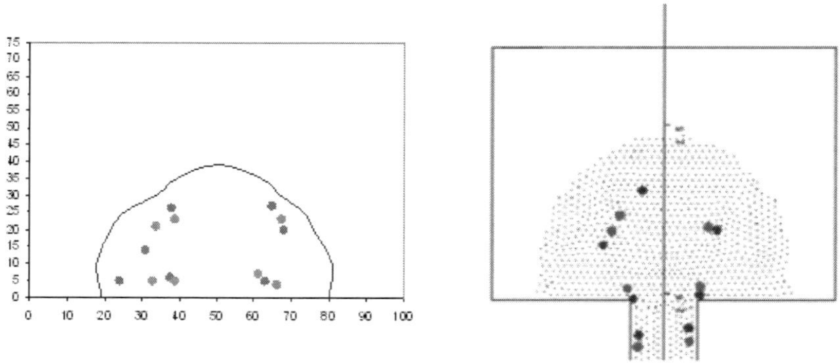

Figure 2: Exp. 1, final position of particles, (a) experimental data (b) numerical solution.

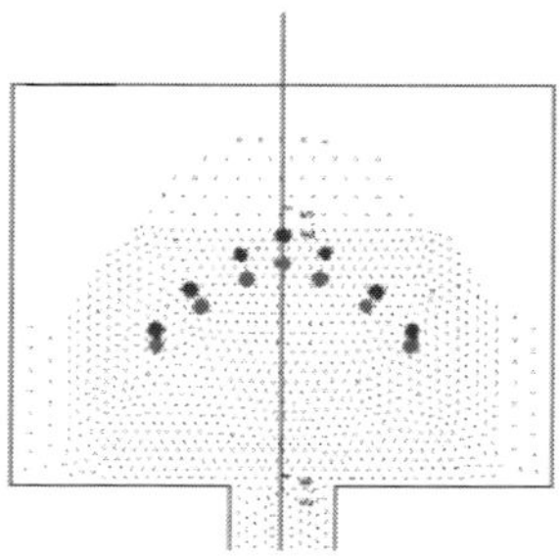

Figure 3: Exp. 1, final position of particles with fully slip boundary condition.

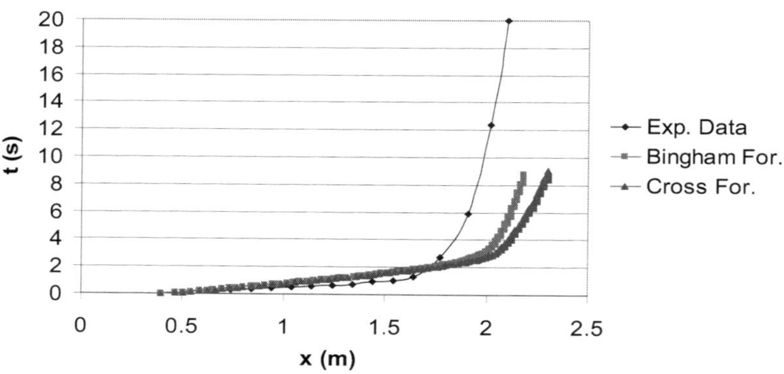

Figure 4: Exp. 2, spreading relation.

Figure 4 shows the spreading relation in the longitudinal direction for this experiment. This relation is compared with numerical results obtained using Bingham rheological model and using Cross rheological model.

Although both rheological formulations produce very similar results, they are not totally capable of resembling the spreading of the flow. However, they show a final fluid extent very similar to the experimental one.

In this experiment 14 particles were placed on the fluid in a similar manner that was done in the previous experiment. In this case, particles depicted the velocity profile shape at early times of the experiment; and as the flow progress down-stream, particles keep the parabolic distribution.

Figure 5 compares the final particle positions obtained numerically against final observed particle location. Note that the model is able to replicate that some particles lag behind close to the flume wall and that the general location of the particles on the alluvial fan is very close to the observed locations.

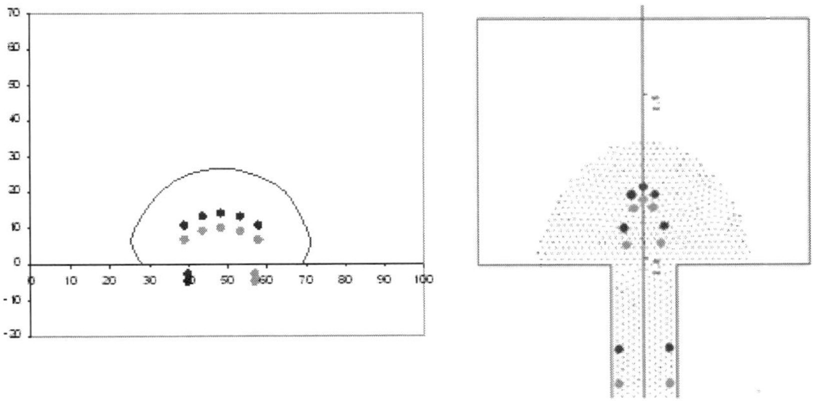

Figure 5: Exp. 2, final position of particles, (a) experimental data (b) numerical sol.

3.3 Model preliminary application: Venezuela's 1999 alluvial fan debris flooding event

Heavy rainfall from a storm on December 14-16, 1999, triggered thousands of shallow landslides on steep slopes of Cerro El Avila, north of Caracas, Venezuela, and caused flooding and massive debris flows in the channels of major drainages that severely damaged coastal communities along the Caribbean Sea. The largest fan on this area is that of San Julián River at Caraballeda, shown in Figure 6. This fan was one of the most heavily damaged areas in the event. The thickness of sediment deposition, maximum size of transported boulders, and size of inundated area were all notably larger in this drainage in comparison to the other close watersheds.

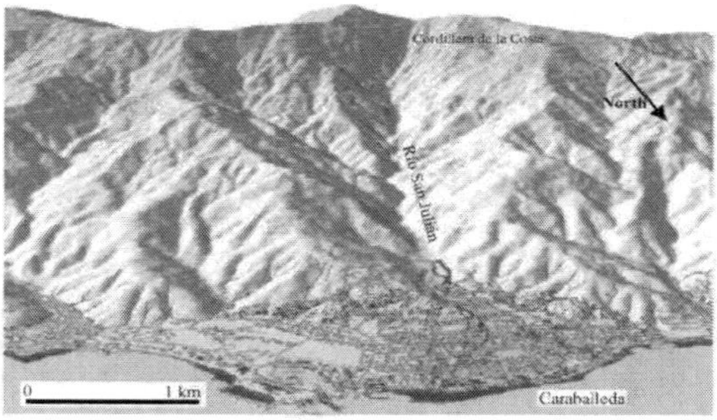

Figure 6: Caraballeda Alluvial Fan, Vargas, Venezuela.

The US Geological Survey studied the affected area [11], measuring slope, deposit thickness, and boulder size from the fan apex to the distal end of the fan near the coastline. Data was used to map the distribution and thickness of deposits and to draw contours of maximum boulder size, as shown in Figure 7.

The numerical simulation was performed using a finite element mesh with 22,500 triangular elements. The element characteristic size was 12 m on average. At the fan apex, a 500 year-return period hydrograph was used as flow input with an average volume sediment concentration of $C_v = 0.3$. Fluid properties are ρ = 1531 Kg/m^3, μ = 0.11 Pa.s, τ_y = 105 Pa. During the simulation, 1600 boulders with sizes ranging from 1 m to 6 m diameter were included in the event. Density for the boulders is ρ = 2600 Kg/m^3, equal to the density of Gneiss boulders, the type of boulders mostly found in the area by the USGS.

Figure 7 shows boulder positions after 6 hours of simulation in comparison with contours of maximum boulder size given by USGS. According to USGS, for station S1 the mean nominal diameter was 1 m, while some larger boulders are deposited slightly further down the fan towards station S2, with 3.5 m nominal diameter. For comparison, boulders deposited at station S3 and S4 had mean nominal diameter of 3 m, and boulders deposited at station S5 had average nominal diameter of 5 m. Figure 7 (b) shows the final distribution of boulders obtained numerically, where it can be seen that the model predicts reasonable boulder locations as compared with the field data.

4 Conclusions

This work describes the development and application of a quasi three-dimensional two-phase model to simulate debris flows, considering large sediment particles, such as boulders. The continuum non-Newtonian phase is solved by the finite element method in 2D and the particle transport with the Discrete Element Method in 3D. The model is able to replicate fluid and particle transport when compared against several experiments in a laboratory flume-fan,

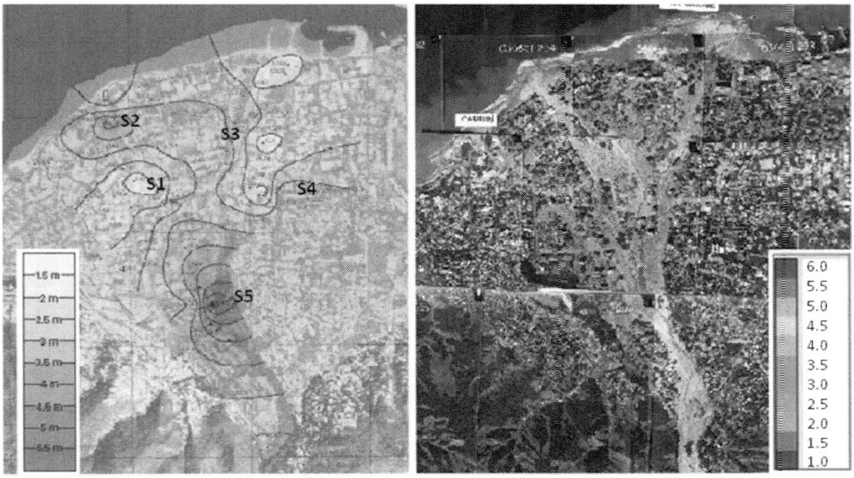

(a) (b)

Figure 7: (a) Contours of maximum boulder size at the fan generated from field data. (b) Particle positions at t = 6.0 h.

including the effect of particle-particle and wall-particle collisions. An application to the well documented debris flow event that occurred in Northern Venezuela in 1999 illustrates the capability of the model to reproduce large scale real events. Results show that the model reasonably approximates the flood extent affected by the debris flow and the observed boulder accumulation areas, including distribution boulders sizes. Future work includes comparison with field events using larger number of boulders to improve predictions.

References

[1] Iverson, R. M. The physics of debris flows. *Rev. of Geophysics*, **35**, pp. 245–296, 1997*b*.

[2] Johnson, A. M. *A model for debris flow*. Ph.D. dissertation. Pennsylvania State University, University Park. 1965.

[3] Bingham, E. C., and Green, H. Paint, a plastic material and not a viscous liquid; the measurement of its mobility and yield value. *Proceedings of American Society of Testing Materials*, **19**, pp. 640-664, 1919.

[4] Barnes, H.A., Hutton J.F., Walters, K. *An introduction to rheology*. Amsterdam. Elsevier. 1989

[5] O'Brien, J.S. and Julien, P.Y. Physical properties and mechanics of hyperconcentrated sediment flows. *ASCE Specialty Conference on the Delineation of Landslides, Debris Flows Hazards*, pp. 260-279, 1985.

[6] Bagnold, R. A. Experiments on a gravity-free dispersion of large solid spheres in a Newtonian fluid under shear. *Proceedings of the Royal Society of London*, **225**, pp. 49-63, 1954.

[7] Takahashi, T. *Debris Flows*. Rotterdam, Balkema. 1991.
[8] Asmar B. N., Langston, P. A. and Ergenzinger, P. The potential of the Discrete Element Method to simulate debris flow. *Debris-flow hazards mitigation: mechanics, prediction and assessment*, **1**, pp. 435-445, 2003.
[9] O'Brien, J.S. and Julien, P.Y. Laboratory analysis of mudflows properties. *J. of Hyd. Eng.*, **114(8),** pp. 877-887, 1988.
[10] García-Martínez, R., Espinoza, R., Valera, E. & González, M. An explicit two-dimensional finite element model to simulate short and long term bed evolution in alluvial rivers. *J. of Hyd. Res.*, **44 (6),** pp. 755-766, 2006.
[11] Wieczorek, G.F., Larsen, M.C., Eaton, L.S., Morgan, B.A. and Blair, J. L. 2001. *Debris-flow and flooding hazards associated with the December 1999 storm in coastal Venezuela and strategies for mitigation. U.S. Geological Survey*, Open File Report 01-0144. 2001.

New formulas for the motion resistance of debris flows

D. Berzi[1], J. T. Jenkins[2] & E. Larcan[1]
[1]*Department of Environmental, Hydraulic, Infrastructures and Surveying Engineering, Politecnico di Milano, Italy*
[2]*School of Civil and Environmental Engineering, Cornell University, USA*

Abstract

We simplify a two-phase theory proposed by Berzi and Jenkins for the uniform motion of a granular-fluid mixture to obtain explicit, analytical relations between the tangent of the angle of inclination of the free surface, the average particle (fluid) velocity and the particle (fluid) depth. Those expressions, valid, in principle, only in uniform flow conditions, can then be employed to express the motion resistance for the particles and the fluid in mathematical models of non-uniform flow, as customary in Hydraulics. The advantages of those formulas with regard to previous, widely employed expressions are also discussed.
Keywords: rheology, uniform flow, friction slope.

1 Introduction

Recently, Berzi and Jenkins [1–3] proposed a simple theory based on a linear rheology for the particle interactions, turbulent shearing of the fluid, buoyancy, and drag. They provided a complete analytical description of the steady, uniform flow of a granular-fluid mixture (debris flow) over an inclined bed contained between frictional sidewalls. In order to obtain such analytical solution, they assumed a constant concentration in the particle-fluid mixture and the similarity of the particle and fluid velocity profiles. The predictions of this description compared favourably with the measurements in experiments on steady, uniform granular-fluid flows performed by Armanini *et al.* [4] and Larcher *et al.* [5] on mono-dispersed plastic cylinders and water. As seen in the experiments, the particle and fluid velocity distributions, the flow depths, and the free surface

WIT Transactions on Engineering Sciences, Vol 67, © 2010 WIT Press
www.witpress.com, ISSN 1743-3533 (on-line)
doi:10.2495/DEB100041

inclination were completely determined by the particle and fluid volume fluxes. Here, we simplify the theory of Berzi and Jenkins [1–3] by neglecting the turbulent shear stress in the mixture and the presence of the sidewalls. We can therefore obtain explicit relations between the average particle velocity, the depth and the tangent of the angle of inclination of free surface and between the average fluid velocity, the depth and the tangent of the angle of inclination of free surface. Those relations can then be used as analytical expressions of the motion resistance encountered by the particles and the fluid, respectively, in a debris flow, by interpreting the angle of inclination of the free surface as the so called friction slope.

The paper is organized as follows: first, we briefly recall the theory of Berzi and Jenkins [1–3]; then, we derive simplified expressions for the friction slopes and, finally, discuss them in comparison with other well-known formulas.

2 Theory

We let ρ denote the fluid mass density, c the particle concentration, g the gravitational acceleration, σ the particle specific mass, d the particle diameter, η the fluid viscosity, U the fluid velocity, and u the particle velocity. The Reynolds number $R = \rho d(gd)^{1/2}/\eta$ characterizes the fall velocity of the particles. In what follows, we phrase the momentum balances and constitutive relations in terms of dimensionless variables, with lengths made dimensionless by d, velocities by $(gd)^{1/2}$, and stresses by $\rho\sigma gd$.

We take $z = 0$ to be the top of the grains, $z = h$ to be the position of the rigid bed, and H to be the height of the water above a bed of inclination θ. The degree of saturation, $\xi = H/h$, is greater than unity in the over-saturated case and less than unity in the under-saturated. Sketches of over- and under-saturated flows are depicted in figure 1, together with a generic velocity profile for the particles.

We assume that it is possible to apply the rheology proposed by the French group GDR MiDi [6]. This rheology provides the particle stress ratio $\mu \equiv s/p$ and the concentration c as unique functions of the inertial parameter

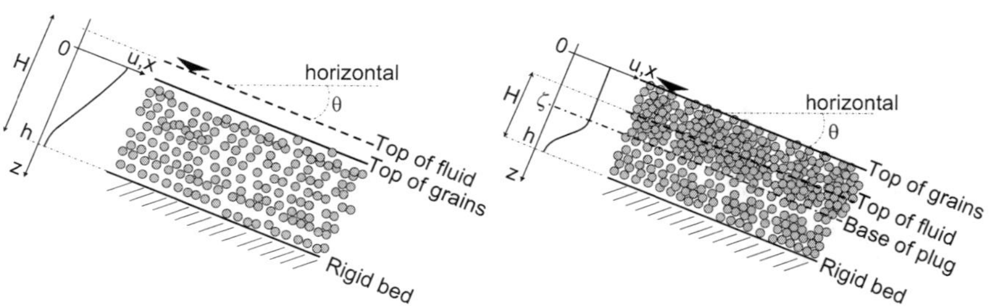

Figure 1: Sketch of steady, (a) over- and (b) under-saturated, uniform flows over rigid beds.

$I \equiv |\dot{\gamma}|/(p/c)^{1/2}$, where s is the particle shear stress, p the particle effective pressure and $\dot{\gamma}$ is the strain rate. In this case, $|\dot{\gamma}| = -u'$; where here, and in what follows, a prime indicates a derivative with respect to z.

We consider highly concentrated flows, in which the functions are approximately linear [7],

$$\mu = \breve{\mu} + \chi I \tag{1}$$

and $c = \widehat{c} - bI$, where $\breve{\mu}$ and $\widehat{c}$ are the minimum stress ratio and the maximum concentration, respectively, and χ and b are material coefficients. The quantities $\breve{\mu}$ and $\widehat{c}$ characterize both the bed and the plug, at which $I = 0$; $\breve{\mu}$ is the tangent of the angle of repose and $\widehat{c}$ is the concentration at dense, random packing.

The balances of fluid momentum transverse and parallel to the flow, in the region in which both phases are present, are

$$P' = \cos\theta / \sigma, \tag{2}$$

and

$$S' = (1-c)\sin\theta / \sigma - cC(U-u)/\sigma, \tag{3}$$

respectively, where P is the fluid pressure, S the fluid shear stress, and C is the dimensionless drag,

$$C \equiv (3|U-u|/10 + 18.3/R)/(1-c)^{3.1}, \tag{4}$$

derived by Dallavalle [8], with the concentration dependence suggested by Richardson and Zaki [9]. When an upper clear fluid layer is present, the distribution of the fluid shear stress can be obtained from eqn. (3) with $c = 0$.

The balances of particle momentum transverse and parallel to the flow are

$$p' = (1 - 1/\sigma)c\cos\theta, \tag{5}$$

and

$$s' = c\sin\theta + cC(U-u)/\sigma, \tag{6}$$

respectively. The balances for the particles when an upper dry layer is present can be obtained from eqns. (5) and (6) by letting σ become infinite.

Here, in the mixture, we ignore the turbulent shear stress in the fluid relative to gravity and drag and neglect the friction of the sidewalls. In the clear fluid layer, we assume that the turbulent mixing length is proportional to the thickness of the layer:

$$S = -k^2 (H-h)^2 \left|U'\right| U', \tag{7}$$

where $k = 0.20$, half the value of Karman's constant. We also assume that the concentration is approximately constant and at its maximum value, $c = \hat{c}$.
With these assumptions, and considering the surface at $z = 0$ as free of particle stress, it is possible to obtain the particle stress ratio, μ, as a function of z from the momentum balances (2), (3), (5) and (6):

$$\mu = \frac{\sigma z + (1-\hat{c})\left[z - h(1-\alpha)\right]/\hat{c}}{\sigma z - z + h(1-\alpha)} \tan\theta + \frac{\sigma S^*}{\left[\sigma z - z + h(1-\alpha)\right]\hat{c}\cos\theta} \tag{8}$$

(for details of this derivation, see [2]), where $\alpha = H/h$ in an under-saturated flow and unity otherwise; and $S^* = h(\beta-1)\sin\theta/\sigma$ is the fluid shear stress at the top of the particles, where $\beta = H/h$ in an over-saturated flow and unity otherwise.

2.1 Particles

In the upper dry layer, μ is constant and equal to $\tan\theta$ (from eqn. 9, with σ equal to infinity). Given the linear rheology (1), in the under-saturated flows, the upper dry layer is either totally sheared, when $\tan\theta > \breve{\mu}$, or there is a plug in the region $\zeta \le z \le 0$. The location ζ of the base of the plug can be found from eqn. (8), with $\mu = \breve{\mu}$ and $S^* = 0$:

$$\frac{\zeta}{h} = \frac{\left[(1-\hat{c})\tan\theta + \hat{c}\breve{\mu}\right](1-\alpha)}{\hat{c}(\sigma-1)(\tan\theta - \breve{\mu}) + \tan\theta}. \tag{9}$$

For reasonable values of $\tan\theta$, eqn. (9) can be approximated by $\zeta/h = 1-\alpha$. In this case, the average particle velocity along h is simply equal to

$$u_A = u_m \alpha + u_{dry}(1-\alpha), \tag{10}$$

where u_m is the mean particle velocity in the mixture layer and u_{dry} the mean particle velocity in the dry layer. The quantity u_m can be obtained once known the velocity distribution in the mixture layer. The latter can be obtained using eqn. (1) in eqn. (9), with $I = -\sigma u' / \left\{\left[(\sigma-1)z + h(1-\alpha)\right]\cos\theta\right\}^{1/2}$, and integrating:

$$\frac{u}{(1-1/\sigma)^{1/2}} = -\frac{2}{3}\frac{(z+L)^{1/2}}{\chi}A\left[(z+L)-3(-N/A+L)\right] + \frac{2}{3}\frac{(h+L)^{1/2}}{\chi}A\left[(h+L)-3(-N/A+L)\right], \tag{11}$$

where $A=\left[\sigma+(1-\breve{c})/\breve{c}\right]\tan\theta/(\sigma-1)-\breve{\mu}$, $L=h(1-\alpha)/(\sigma-1)$, and $N=\left[-(1-\breve{c})\tan\theta/\breve{c}-\breve{\mu}\right]L+h(\beta-1)\tan\theta/(\breve{c}\sigma-\breve{c})$. In obtaining eqn. (11), we have assumed a mild slope, so that $\cos\theta\approx 1$, and a zero slip velocity at the bed. It is then possible to obtain αu_m by integrating eqn. (11) between $(1-\alpha)h$ and h:

$$\alpha u_m = \frac{2}{3}\frac{(\sigma-1)^{1/2}h^{3/2}}{\sigma^{1/2}\chi}\left\{\left[\left(\frac{\sigma-\alpha}{\sigma-1}\right)^{3/2}\alpha-\frac{2}{5}\left(\frac{\sigma-\alpha}{\sigma-1}\right)^{5/2}+\frac{2}{5}(1-\alpha)^{5/2}\left(\frac{\sigma}{\sigma-1}\right)^{5/2}\right]A + \left[3\left(\frac{\sigma-\alpha}{\sigma-1}\right)^{1/2}\alpha-2\left(\frac{\sigma-\alpha}{\sigma-1}\right)^{3/2}+2(1-\alpha)^{3/2}\left(\frac{\sigma}{\sigma-1}\right)^{3/2}\right] \times\left(\frac{N}{h}-\frac{1-\alpha}{\sigma-1}A\right)\right\}. \tag{12}$$

If, in the upper dry layer, there is a plug ($\tan\theta\le\breve{\mu}$), its velocity is equal to the velocity u_ζ at the top of the mixture. If the upper dry layer is sheared ($\tan\theta>\breve{\mu}$), from eqn. (1) and the fact that, in the dry layer, $\mu=\tan\theta$ and $I=-u'/z^{1/2}$, the velocity there is equal to

$$u = u_\zeta + \frac{2}{3}\frac{\tan\theta-\breve{\mu}}{\chi}\left(\zeta^{3/2}-z^{3/2}\right). \tag{13}$$

The quantity $(1-\alpha)u_{dry}$ is, then, equal to

$$(1-\alpha)u_{dry} = \begin{cases} (1-\alpha)u_\zeta & \textit{if } \tan\theta\le\breve{\mu} \\ (1-\alpha)u_\zeta+\frac{2}{5}\frac{h^{3/2}}{\chi}(1-\alpha)^{5/2}(\tan\theta-\breve{\mu}) & \textit{if } \tan\theta>\breve{\mu} \end{cases}, \tag{14}$$

where u_ζ can be obtained from eqn. (11) with $z=(1-\alpha)h$. With this and eqns. (12) and (14), eqn. (10) may be written as

$$\frac{u_A}{h^{3/2}} = \lambda_1\tan\theta-\lambda_2, \tag{15}$$

where the coefficients λ_1 and λ_2 are functions of the type of fluid and granular material (through σ, χ $\breve{\mu}$ and $\hat{c}$) and the degree of saturation (through α and β); expressions for them are given in table 1. In uniform flows, the friction slope is equal to the tangent of the angle of inclination of the free surface. An expression for the friction slope, j, for the particles to be used also in non-uniform flows, can, therefore, be obtained from eqn. (15), with $j = \tan\theta$,

$$j = \frac{1}{\lambda_1} \frac{u_A}{h^{3/2}} + \frac{\lambda_2}{\lambda_1}. \tag{16}$$

Table 1: Values of the coefficients in the flow rule for the particles (eqn. (15)).

λ_1 when $\tan\theta \le \breve{\mu}$	$\frac{2}{15\chi\hat{c}\sigma^{1/2}(\sigma-1)^3}\left\{\left[(3\sigma-5+2\alpha)(\sigma-\alpha)^{3/2}-\sigma^{3/2}(3\sigma-5)(1-\alpha)^{5/2}\right](\hat{c}\sigma+1-\hat{c}) +5\left[(\sigma-3+2\alpha)(\sigma-\alpha)^{1/2}-\sigma^{1/2}(\sigma-3)(1-\alpha)^{3/2}\right]\left[(\sigma-1)(\beta-1)-\sigma(1-\alpha)\right]\right\}$
λ_1 when $\tan\theta > \breve{\mu}$	$\frac{2}{15\chi\hat{c}\sigma^{1/2}(\sigma-1)^3}\left\{\left[(3\sigma-5+2\alpha)(\sigma-\alpha)^{3/2}-\sigma^{3/2}(3\sigma-5)(1-\alpha)^{5/2}\right](\hat{c}\sigma+1-\hat{c}) +5\left[(\sigma-3+2\alpha)(\sigma-\alpha)^{1/2}-\sigma^{1/2}(\sigma-3)(1-\alpha)^{3/2}\right]\left[(\sigma-1)(\beta-1)-\sigma(1-\alpha)\right] +3\hat{c}\sigma^{1/2}(\sigma-1)^3(1-\alpha)^{5/2}\right\}$
λ_2 when $\tan\theta \le \breve{\mu}$	$\frac{2}{15\chi\sigma^{1/2}(\sigma-1)^2}\left[(3\sigma-5+2\alpha)(\sigma-\alpha)^{3/2}-\sigma^{3/2}(3\sigma-5)(1-\alpha)^{5/2}\right]\breve{\mu}$
λ_2 when $\tan\theta > \breve{\mu}$	$\frac{2}{15\chi\sigma^{1/2}(\sigma-1)^2}\left[(3\sigma-5+2\alpha)(\sigma-\alpha)^{3/2}-\sigma^{3/2}(3\sigma-5)(1-\alpha)^{5/2} +3\sigma^{1/2}(\sigma-1)^2(1-\alpha)^{5/2}\right]\breve{\mu}$

2.2 Fluid

The average fluid velocity along H is equal to

$$U_A = \frac{\alpha(1-\hat{c})U_m + (\beta-1)U_{cm}}{\alpha(1-\hat{c})+\beta-1}, \tag{17}$$

where U_m and U_{cm} are the mean fluid velocities in the mixture and in the upper clear fluid layer, respectively. Berzi and Jenkins [1, 2] have shown that the calculated difference between the fluid and the particle velocity is rather small (however, this does not permit the neglect of the drag force in the momentum balances (3) and (6), given the high values of the drag coefficient C). We can, therefore, assume that $U_m \approx u_m$.

The mean fluid velocity in the upper clear fluid layer can be obtained from the integration of the distribution of the fluid velocity there; the latter comes from the integration of eqn. (7), with the distribution of the fluid shear stress provided by eqn. (3) when $c = 0$. Hence,

$$U_{cm}(\beta-1) = U_0(\beta-1) + \frac{2(\beta-1)^{3/2}}{5k} h^{1/2}(\tan\theta)^{1/2}, \tag{18}$$

where U_0 is the fluid velocity at the base of the upper clear fluid layer, which can be obtained from eqn. (11) with $z = 0$. With this and eqns. (12) and (18), eqn. (17) reads

$$\frac{U_A}{H^{3/2}} = \Lambda_1 \tan\theta - \Lambda_2 + \frac{\Lambda_3}{H}(\tan\theta)^{1/2}, \tag{19}$$

where Λ_1, Λ_2 and Λ_3 are functions of the type of fluid and granular material (through σ, χ $\breve{\mu}$ and $\hat{c}$), the mixing length (through k), and the degree of saturation (through α and β), and their expressions are given in table 2. Once again, in uniform flows, the friction slope is equal to the tangent of the angle of inclination of the free surface. An expression for the friction slope, J, for the fluid, to be used also in non-uniform flows, can, therefore, be obtained from eqn. (19), with $J = \tan\theta$,

$$J = \left\{ \frac{-\Lambda_3 + \left[\Lambda_3^2 + 4\Lambda_1\left(\Lambda_2 H^2 + U_A H^{1/2}\right)\right]^{1/2}}{2\Lambda_1 H} \right\}^2. \tag{20}$$

3 Discussion

We have simplified the theory proposed by Berzi and Jenkins [1–3] to obtain explicit relations between the tangent of the angle of inclination of the free

Table 2: Values of the coefficients in the flow rule for the fluid (eqn. (19)).

Λ_1	$\frac{2(1-\hat{c})}{15\xi^{3/2}\chi\hat{c}\sigma^{1/2}(\sigma-1)^3\left[\alpha(1-\hat{c})+\beta-1\right]}\left\{\left[(5\alpha\sigma-3\alpha-2\sigma)(\sigma-\alpha)^{3/2}+2\sigma^{5/2}(1-\alpha)^{5/2} +5(\sigma-1)^{5/2}(\beta-1)\right](\hat{c}\sigma+1-\hat{c}) +5\left[(3\alpha\sigma-\alpha-2\sigma)(\sigma-\alpha)^{1/2}+2\sigma^{3/2}(1-\alpha)^{3/2}\right]\left[(\sigma-1)(\beta-1)-\sigma(1-\alpha)\right] +15(\beta-1)^2(\sigma-1)^{5/2}\right\}$
Λ_2	$\frac{2(1-\hat{c})}{15\xi^{3/2}\chi\sigma^{1/2}(\sigma-1)^2\left[\alpha(1-\hat{c})+\beta-1\right]}\left[(5\alpha\sigma-3\alpha-2\sigma)(\sigma-\alpha)^{3/2}+2\sigma^{5/2}(1-\alpha)^{5/2}+5(\sigma-1)^{5/2}\right.$
Λ_3	$\frac{2(\beta-1)^{3/2}}{5\xi^{1/2}k\left[\alpha(1-\hat{c})+\beta-1\right]}$

surface, the depth and the average particle velocity and between the tangent of the angle of inclination of the free surface, the depth and the average fluid velocity. Those two relations are the flow rules for the particles and the fluid, respectively, if one interprets the tangent of the angle of inclination of the free surface as the friction slope.

The fact that the friction slope for the particles has a different expression from that for the fluid is crucial to the expression of the resistances in two-phase, depth-averaged, mathematical models of non-uniform flows (see, for example, the steady granular-fluid wave over a rigid bed analysed in [3]). Most previous works either treat the mixture as a single phase fluid [10–14] or, although aware of the differences between the two phases, focus solely on the particle motion resistance [15, 16].

Existing models for the motion resistance of debris flows can be basically grouped in the four categories described in the following (although there are examples of resistance formula obtained by combining the characteristics of two categories, e.g. see [17]); however, all of them suffer from major drawbacks with respect to the formulas presented here.

Takahashi [15] obtains an expression for the resistance of over-saturated debris flows, based on a modified version of the dilatant model for the particle shear stresses in the inertial regime described by Bagnold [18] using kinetic arguments. Certainly, the merits of Takahashi expression were his taking into account the dependence of the stress ratio on the particle concentration and his incorporation of the effects of the fluid turbulence. However, his theory was incomplete, because it did not deal with under-saturated debris flows and because he characterized the particles only through their density.

Some authors [16, 19] suggest the use of Coulomb's law to express the friction at the base of a debris flow. However, Coulomb's law cannot explain the experimentally observed dependence of the friction slope on the average velocity and the depth [20], given that it implies a constant stress ratio at the bed. In the theory of Berzi and Jenkins [1–3], the stress ratio at the bed depends on the local inertial parameter, i.e. the velocity gradient.

Many authors employ some kind of non-Newtonian rheology for modelling the debris flow resistance [10–14]. This approach implies that the debris flow can be approximated as a single-phase fluid. This, perhaps, applies when the solid phase is composed mainly of fine sediments (e.g. for mud flows, see [4] for more details) - that is, when the inertia of the particles is negligible with respect to the fluid viscous forces; but not when the content of large particles is relevant (as for stony debris flows, see [4]). The assumed non-Newtonian behaviour of the debris flow is, moreover, entirely phenomenological and, therefore, not well physically-based. Although the GDR MiDi rheology adopted here might also seem phenomenological, its physical link with the particle interactions at the micromechanical level has been demonstrated [21].

Finally, a few authors [14, 17] employ empirical expressions for the friction slope based on that for purely turbulent fluids (the Manning equation). These are not physically based and there are no rational arguments to justify their usage.

The formulas for the motion resistance of particles and fluid in debris flows proposed in the present work seem promising for practical application in the field of civil engineering.

References

[1] Berzi D. & Jenkins J.T., A theoretical analysis of free-surface flows of saturated granular-liquid mixtures. *J. Fluid Mech.*, **608**, pp. 393–410, 2008.

[2] Berzi D. & Jenkins J.T., Approximate analytical solutions in a model for highly concentrated granular-fluid flows. *Phys. Rev.* E, **78**, pp. 011304, 2008.

[3] Berzi D. & Jenkins J.T., Steady inclined flows of granular-fluid mixtures. *J. Fluid Mech.*, **641**, pp. 359–387, 2009.

[4] Armanini, A., Capart, H., Fraccarollo, L. & Larcher, M., Rheological stratification in experimental free-surface flows of granular-liquid mixtures. *J. Fluid Mech.,* **532**, pp. 269–319, 2005.

[5] Larcher, M., Fraccarollo, L., Armanini, A. & Capart, H., Set of measurement data from flume experiments on steady, uniform debris flows. *J. Hydr. Res.*, **45**, pp. 59–71, 2007.

[6] GDR MiDi, On dense granular flows. *Eur. Phys. J.* E, **14**, pp. 341–365, 2004.

[7] da Cruz, F., Sacha, E., Prochnow, M., Roux, J. & Chevoir, F., Rheophysics of dense granular materials: Discrete simulation of plane shear flows. *Phys. Rev.* E, **72**, pp. 021309, 2005.

[8] Dallavalle, J., *Micromeritics*, Pitman: New York, 1943.

[9] Richardson, J.F. & Zaki, W.N., Sedimentation and fluidization. *Trans. Inst. Chem. Engrs.*, **32**, pp. 35–53, 1954.

[10] Coussot, P., Steady, laminar flow of concentrated mud suspensions in open channel. *J. Hydraul. Res.*, **32(4)**, pp. 535–559, 1994.

[11] Chen, C.L. & Ling, C.H., Rheological equations in asymptotic regimes of granular flow. *J. Eng. Mech.-ASCE*, **124(3)**, pp. 301–310, 1998.

[12] O'Brien, J.S., Julien, P.Y. & Fullerton, W.T., Two-Dimensional Water Flood and Mudflow Simulation. *J. Hydraul. Eng.-ASCE*, **119(2)**, pp. 244–261, 1993.

[13] Brufau, P., Garcia-Navarro, P., Ghilardi, P., Natale, L. & Savi, F., 1-D Mathematical modelling of debris flow. *J. Hydraul. Res.*, **38**, pp. 435–446, 2000.

[14] Berzi, D. & Larcan, E., Transient hyper-concentrated flows: limits of some hypotheses in mathematical modeling. *Proc. of the 2nd Int. Conf. on Fluvial Hydraulics River Flow 2004*, ed. M. Greco, Taylor & Francis Ltd., pp. 1103–1110, 2004.

[15] Takahashi, T., *Debris flow. IAHR Monograph Series*, Balkema, 1991.

[16] Iverson, R.M., The physics of debris flows. *Rev. Geophys.*, **35**, pp. 245–296, 1997.

[17] Hungr, O., A model for the runout analysis of rapid flow slides, debris flows, and avalanches. *Can. Geotech. J.*, **32**, pp. 610–623, 1995.

[18] Bagnold, R.A., Experiments on a gravity-free dispersion of large solid spheres in a Newtonian fluid under shear. *Proc. R. Soc. London A*, **225**, pp. 49–63, 1954.
[19] Pitman, E.B. & Le, L., A two-fluid model for avalanche and debris flows. *Phil. Trans. R. Soc. A*, **363**, pp. 1573–1601, 2005.
[20] Ancey, C. & Evesque, P., Frictional-collisional regime for granular suspension flows down an inclined channel. *Phys. Rev.* E, **62**, pp. 8349–8360, 2000.
[21] Jenkins, J.T., Dense inclined flows of inelastic spheres. *Gran. Matter*, **10**, pp. 47–52, 2007.

Rheological behaviour of pyroclastic debris flow

A. M. Pellegrino[1], A. Scotto di Santolo[1], A. Evangelista[1] & P. Coussot[2]
[1]*Department of Hydraulic, Geotechnical and Environmental Engineering, University of Naples "Federico II", Italy*
[2]*Université Paris – Est, Institut Navier, France*

Abstract

The pyroclastic soils that cover the mountains of the Campania region in Italy are usually unsaturated and collapse due to rainfall infiltration triggering landslides. The evolution of these soils after collapse is not well understood. Indeed, their post-failure behaviour may be "solid-like" or "fluid-like", depending on causes that are not well known. The objective of this paper is the study carried out on the rheological behaviour of the "fluid-like" pyroclastic material with fluid mechanics tools: a vane rotor rheometer and an inclined plane. Two natural pyroclastic deposits have been sampled and different soils-water mixtures have been analysed. The main results have been explained and discussed
Keywords: debris flow, pyroclastic soil, solid-liquid transition, rheology, laboratory activity, fluids model, yield stress, critical shear rate.

1 Introduction

The Campania region has been covered by pyroclastic deposits generated by different volcanic centres, the most famous of which is the Somma-Vesuvius, which is still active inside the so-called Campanian Volcanic Zone. In this area, pyroclastic soils (mostly ash and pumices) and soft rocks (tuff) have been extensively used since antiquity for construction purposes. The cover is cohesionless and poses severe slope stability problems. The landslides have been classified as translational or rotational sliding or falls that lead to debris flows. As a result of the ceaseless growth and spreading of urbanised areas and

WIT Transactions on Engineering Sciences, Vol 67, © 2010 WIT Press
www.witpress.com, ISSN 1743-3533 (on-line)
doi:10.2495/DEB100051

infrastructures, the risk of landslides have been increased enormously, as testified by hundreds of victims of flowslides in the last fifty years [1, 2].

Despite the relevance of the problem, a comprehensive geotechnical classification of these deposits is still lacking. While the mechanical properties of the Campanian natural soils are the object of numerous studies, the post-failure behaviour of the material may be “solid-like” or “fluid-like” according to causes that are not well known. This paper has been focused on the post-failure behaviour of such materials which it seems to depend on several factors (i.e., the geometry of the slope, the sequence and mechanical properties of the cover, the porosity, the grain size distribution, the stress conditions and the characteristics of the rainfall prior to and following the triggering) [3, 4].

As an idealisation, a debris flow has been often considered as a mixture of viscous slurry, made up of the finer grain sizes and water, and coarse particles. In general, fluid-to-fluid, fluid-to-solid, and solid-to-solid interactions can play an important role [5]. Some researchers have been used such codes for back-analysing debris flows in the Campania region [6–8]. Alternatively, the use of a fluid mechanical treatments to study the rheological behaviour of these pyroclastic deposits remixed with water has not been studied often because of the flow curve (the relationship between shear stress and shear rate in the steady state) is nevertheless not so easy to evaluate. Scotto di Santolo [3, 4] and Scotto di Santolo et al. [9] have carried out preliminary rheometrical tests on fine particle-water suspensions of some pyroclastic soils collected in Campania region at different solid volumetric concentrations; they have found that the mixture behaves as a non-Newtonian fluid with a yield stress and that the behaviour varying with the solid concentration.

In this paper, the further results of a more complete laboratory activity on two of these pyroclastic deposits have been reported in order to evaluate whether the rheological approach can be a useful tool for understanding in which condition the pyroclastic soil changes behaviour from that of a soil to that of a fluid.

2 Materials

The materials tested have been collected from the source area of two debris flows in the Campania region [9]. Material I has been sampled in Nocera, Salerno (March 2005) and material II has been sampled in Monteforte Irpino, Avellino (May 1998). The soil type, in a thickness of about a metre, depends on the most recent pyroclastic deposits deriving from the volcanic activity of Mount Somma/Vesuvius [3, 4, 10]. The main grain size distributions of the collected samples are reported in fig. 1. Soil I and soil II are sandy silt with a small clay fraction. The bedrock underlying the soil is limestone for materials I and II. Mean physical properties are reported in Table 1 (where G_S is the specific gravity of soil particles, γ_d and γ are the dry and total weight of soil per unit volume respectively, n is the porosity, and S_r is the degree of saturation).

Due to the size of the rheometrical facilities [11], the soil fraction with a particle diameter less than 0.5 mm has been kept. This represents about 70% of the whole grain size distribution (see fig. 1), so it has been expected that the behaviour of this material represents the behaviour of the full mixture well.

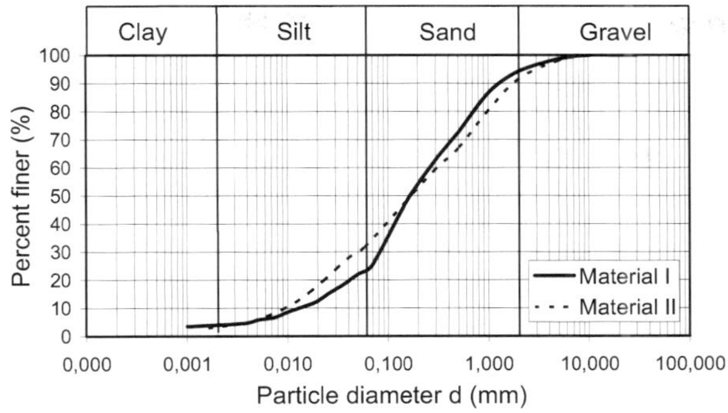

Figure 1: Grain size distribution of the natural deposits analysed.

Table 1: Main physical properties of the tested debris flow materials.

Debris flow Site	Substratum	Material	G_S (1)	γ_d (kN/m^3)	γ (kN/m^3)	n (1)	S_r (1)
Nocera (SA)	Carbonatic	I	2.61	9.08	11.35	0.66	0.35
Monteforte Irpino (AV)	Carbonatic	II	2.57	7.11	12.11	0.71	0.71

All experiments have been carried out with mixtures of different water contents in order to consider a significant range of the sediment concentrations for each material tested (according to the in situ porosity; see Table 1). The solid volumetric concentration, i.e., the ratio of the volume of solids to the total volume (water plus solids), has been used and it has been defined as:

$$\Phi = \frac{V_s}{V_w + V_S} \tag{1}$$

where V_w and V_s are, respectively, the volumes of water and solid in the sample. For each material tested, material mixtures of about 500 ml have been prepared, mixing soils and water with an electronic mixer for 15 minutes. Then a sample volume of about 30 ml for each test has been used at a constant temperature (23°C).

3 Set up and procedures

The behaviour of the material mixtures analysed has been investigated with two experimental apparatuses: a conventional rotational rheometer and an inclined plane.

3.1 Rheometer

A rotational rheometer CVOR (Bohlin Instruments) equipped with a vane rotor geometry system (fig. 5a) has been used. It consists of four thin blades arranged at equal angles around a small cylindrical shaft: the blade radius was 13 mm, and the blade height was 48 mm. The vane rotor has been immersed in the sample contained in a cylindrical cup 18.5 mm in radius. The rotor has been rotated around its axis at a given rotational speed Ω, and the torque T has been measured. Under usual assumptions (no inertia effects, negligible normal stress differences) the shear stress and the shear rate within the material are given as:

$$\dot{\gamma} = \frac{\Omega \quad R_1}{R_2 \quad R_1} \tag{2}$$

$$\tau = \frac{T}{2\,\pi\,R_1^2 L} \tag{3}$$

where R_1 and R_2 are, respectively, the blade radius and the cup radius, and L is the material depth. Two kinds of tests have been carried out: creep and stress sweep. During the creep test, a constant torque (associated with a constant stress) has been imposed and the material behaviour has been monitored from the resulting deformation versus time curve. Deformation is expressed in terms of angle of creep (i.e., the angle, in radians, of the displacement after the creep stress was applied). During a stress sweep, the flow curves have been determined by applying an increasing-decreasing shear stress ramp. In that case, the material response has been followed from the resulting shear stress – shear rate curve. Before each test, a pre-shear has been applied, which consists of imposing a very rapid flow and then leaving the material at rest for some time in order to provide a homogeneous state of the sample before carrying out each test defined.

3.2 Inclined plane

The inclined plane test consists of leaving a certain amount of paste over an inclined plane and analysing the fluid depth profile in rheological terms. The equipment for the test consists of a roughness plane and an inclinometer (fig. 2b). The test procedure is the following: first the mixtures have been spilled on

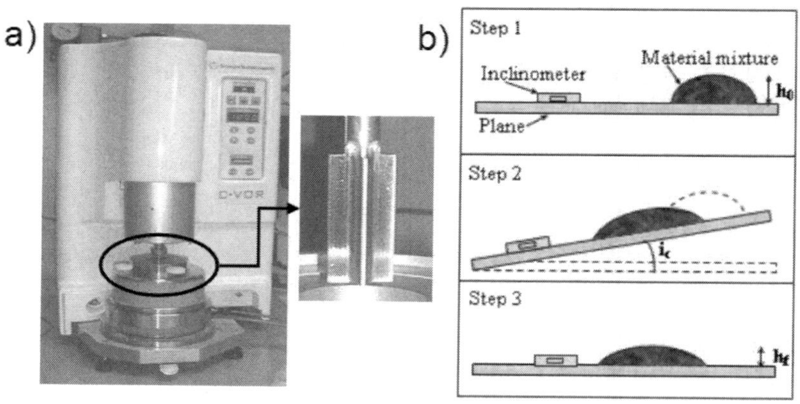

Figure 2: Experimental apparatus used: a) the rotational rheometer (CVOR Bohlin) and the vane rotor geometric system; b) scheme of the inclined plane test procedure.

the plane (step 1 in fig. 2b) and with a graduated rule the initial thickness of the deposits has been measured, h_0, as the averaged value of several measurements. Then, the plane has been inclined and stopped it when the front of the deposit started to move (step 2 in fig. 2b). At that moment, the critical angle has been defined, i_c, as the corresponding inclination of the plane. Finally the final thickness of the deposits, h_f, has been defined as the averaged value (step 3 in fig. 2b). Under the usual so-called "lubrication assumption" [12], a simple momentum balance provides the shear stress distribution within the material and the critical thickness may be used for determining two critical stresses: a "static" yield stress (τ_{c1}) and a "dynamic" yield stress (τ_{c2}):

$$\tau_{C1} = \rho g h_0 \sin(i_c) \quad (4)$$

$$\tau_{C2} = \rho g h_f \sin(i_c) \quad (5)$$

where ρ is the mixture density and g is the acceleration of gravity.

4 Behaviour evolution with the solid fraction

The first step with the selected materials was to investigate their overall mechanical behaviour as a function of the solid concentration. By mixing the solid with water, a mixture has been obtained that was in three possible states:

- For sufficiently low volume fractions, the solid particles rapidly (within a few seconds) settle down, leading to an apparent phase separation; such a situation follows from the fact that the suspended particles do not interact when they are dispersed in water (such as in the very first seconds after preparation) so that we are dealing with a Newtonian fluid with an apparent viscosity typically of the order of ten times that of pure water; when the particles have settled, we are no longer dealing with a homogeneous material, and nothing can be said about its viscosity.
- For too high volume fractions, the suspension obtained is in fact a kind of paste of high strength, which easily breaks like a solid when it is deformed; such a material cannot be considered as a fluid able to undergo reversible large deformations without changing its basic properties.
- For intermediate volume fractions, we can observe some slight sedimentation after significantly longer times of rest, a point that we will discuss below; the material thus remains homogeneous over a reasonable time of observation and can flow like a liquid.

The lower bound is 32% for material I and 30% for material II. These values were very low for each material tested, slightly different from each other but in agreement with the porosity of the natural deposits (shown in table 1). The upper bound of solid concentration is 42% for material I and 38% for material II. For higher solid volumetric concentrations, no steady flow regime occurred.

Surprisingly, the range of solid concentrations in which the material mixtures can be considered as "fluid-like" is rather narrow for each material tested, in contrast with clay-water systems, for example, for which one may get homogeneous fluids in a range from one percent to several tens of percents.

Subsequently the "fluid-like" behaviour within these intervals identified for each material has been investigated.

5 Rheometer results

5.1 Creep test results

The results of creep tests conducted on material I mixtures at two different volumetric solid concentrations are reported in fig. 3. For stresses lower than the critical value the curve remains concave with a slope continuously decreasing in time and exhibits an apparent horizontal asymptote: no steady state has been reached and the material apparently stops moving. This should correspond to the solid regime of the material. For stress values higher than the critical value the initial slope of the curve is similar to that under smaller stresses, but after some time, there is an inflection point, and the curves tend to reach an inclined straight line gradually with a slope equal to 1: a steady state flow has been reached. These curves correspond to the liquid regime of the material: the deformation trend is linear with time and the material shows a constant shear rate when a constant shear stress is applied. In this context, this critical value of stress has been defined the static yield stress, τ_{c1}, which is the value of stress at which the material ultimately flows in a liquid regime. When looking at the set of creep curves for different stress values, different aspects have been seen. In some cases (see fig. 3a), there is apparently a smooth transition from the solid to the liquid regime. Such behaviour corresponds to simple yield stress behaviour: around the yield stress, the material changes from a situation in which it does not flow (solid regime) to a situation in which it flows extremely slowly (liquid regime just above the yield stress). In other cases, the transition from the solid to the liquid regime is more abrupt (see fig. 3b): around the yield stress the material changes from not flowing to flowing at a relatively high shear rate.

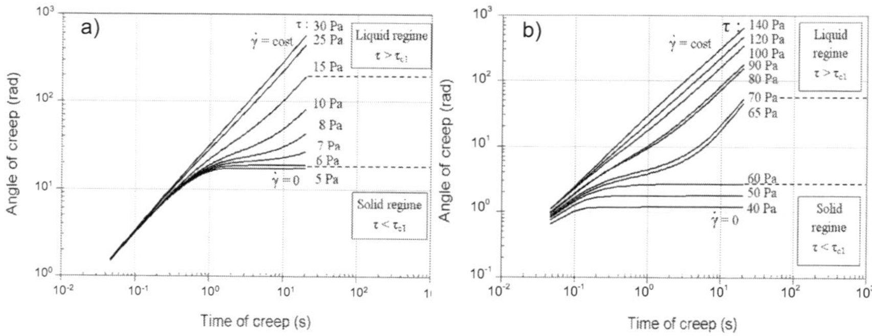

Figure 3: Creep curves for different stress values of material I at: a) solid concentration of 35% and b) solid concentration of 40%.

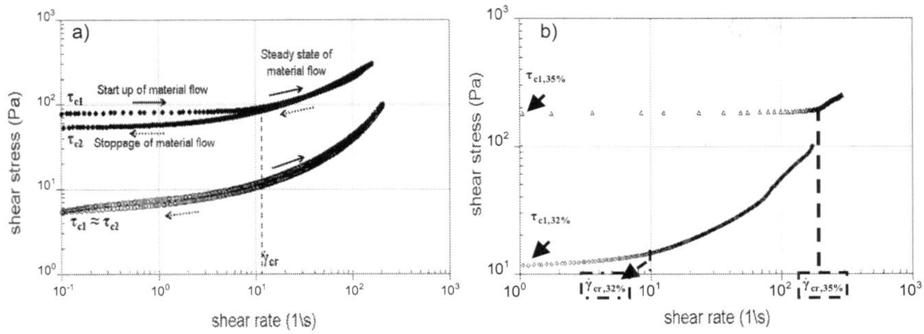

Figure 4: a) Flow curves of material I mixtures at two different solid concentrations obtained with an increasing stress ramp (arrow with continuous line) and a decreasing stress ramp (arrow with dashed line). The filled circles are the flow curves at a solid concentration equal to 40%, while the empty circles are the flow curves at a solid concentration equal to 35%. b) Increasing the part of flow curves of material II mixtures at two different solid concentrations. The empty circles are the flow curve at a solid concentration equal to 32%, while the empty triangles are the flow curve at a solid concentration equal to 35%. The dashed line is associated with the critical value of the shear rate for each material mixture tested.

5.2 Sweep test results

Fig. 4a shows the experimental results for the flow curves of two mixtures of material I at different solid concentrations. Once again, two different trends have been observed. In some cases (fig. 4a for 35% solid concentration), there is first an increase in the stress with the shear rate at low shear stresses. This in fact corresponds to the response of the material in the solid regime. The rest of the flow curve is supposed to correspond to the material behaviour in the liquid regime: the transition to the liquid regime is associated with the rapid increase of the shear rate (stress plateau) above some critical value of the stress. At larger stresses, the curve slope increases. The plateau thus obtained for the stress increase is associated with the static yield stress, τ_{c1}, of the material (as observed in creep tests). Then the decreasing curve falls along the increasing curve. In other cases (fig. 4a for 40% solid concentration), the decreasing curve differs significantly from the increasing curve; there is a hysteresis where the stress for flow in the decreasing stress part is below the increasing curve. One can consider that the material was initially broken then liquefied, and the stress needed to maintain the flow is lower than the stress to break the initial structure. Under these conditions, the dynamic stress, τ_{c2}, has been defined as the critical stress for flow stoppage when decreasing the stress level.

Generally, a material exhibiting a smooth solid-liquid transition (for example, material I at 35% solid concentration, fig. 4a) has a good superposition of the increasing and decreasing stress curves, whereas those exhibiting an abrupt transition (for example, material I at 40% solid concentration, fig. 4a) also exhibit the hysteresis in the flow curves. In the case of a smooth transition, the static yield stress is equal to the dynamic yield stress.

As mentioned above, the material mixtures tested exhibit a stress plateau in the flow curve at a particular stress value (i.e., the static yield stress), and the stress remains approximately constant in a certain range of shear rates. A typical example is shown in fig. 4b, which shows the flow curves of two mixtures of material II at different solid concentrations (32% and 35%) obtained with an increasing ramp of stress. It has been noted that when progressively increasing the stress level, a large increase of the resulting shear rate is observed around a critical value (the static yield stress), which rapidly transforms from a small value to a much larger value associated with the end of the plateau. This last value of the shear rate, called the critical shear rate, $\dot{\gamma}_{cr}$, marks the transition of the material mixture behaviour from a yielding behaviour (low to high shear rate) to a steady state flow (high shear rate). In fact, no steady flows can be obtained below the critical shear rate [12, 13].

Considering that the liquid regime corresponds only to the decreasing stress curve, the Herschel–Bulkley rheological model has been used to fit it, which is a generalised model of a non-Newtonian fluid valid for yield stress fluids. The eqn. of the Herschel–Bulkley model is:

$$\tau = \tau_{C2} + k\dot{\gamma}^{n} \tag{6}$$

where $\dot{\gamma}$ is the shear rate, k is the consistent coefficient, and n is the flow index.

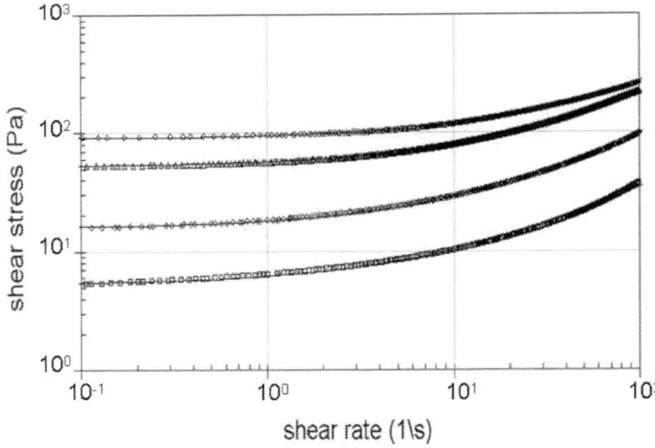

Figure 5: The experimental data (points) and the theoretical Herschel and Bulkley model (solid line) of material I at different solid volumetric concentration: the empty rectangles are the flow curves obtained at solid concentration of 35%; the empty rhombuses are the flow curves obtained at solid concentration of 38%; the empty triangles are the flow curves obtained at solid concentration of 40%; and the empty circles are the flow curves obtained at solid concentration of 42%.

The consistency k is a simple constant of proportionality [Pa·s^n]. The dimensionless index n measures the degree to which the fluid is shear-thinning or shear-thickening [14].

In the ranges of solid concentration in which the materials mixtures analysed could be considered as homogeneous (32-42% for material I and 30-38% for material II), the debris flow material mixtures analysed behave like yield stress fluids. Fig. 6 shows the variation of the rheological parameters with the solid fraction. As mentioned before, it was found that the static yield stress (fig. 6a), the dynamic yield stress (fig. 6b) and the critical shear rate (fig. 6c) are higher with increasing solid fraction. An exponential function could be used to relate the rheological parameters to the solid volumetric concentration:

$$\tau_{c1}, \tau_{C2}, \dot{\gamma}_{cr} = \alpha e^{\beta\Phi} \tag{7}$$

where α and β are fitting parameters, and their values are reported in table 2.

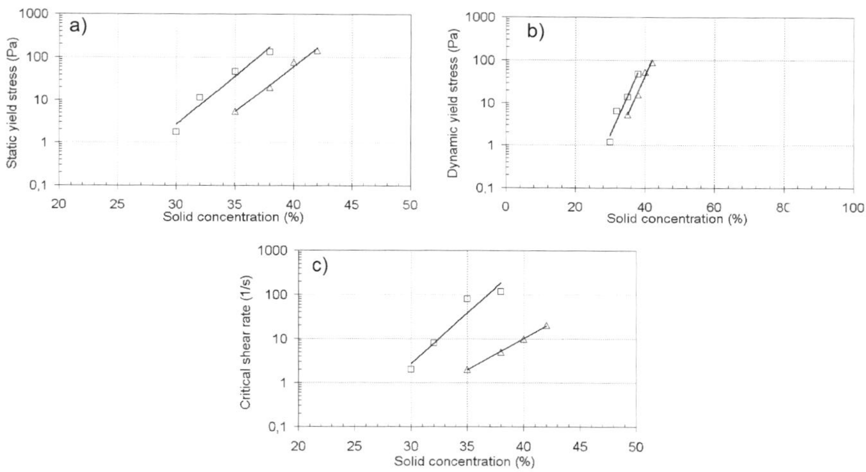

Figure 6: Influence of the solid concentration for the materials tested on: a) the static yield stress; b) the dynamic yield stress and c) the critical shear rate. The triangles are relative to material I and the rectangles are relative to material II.

Table 2: Fitting parameters α and β.

	Material	α	β	R^2
τ_{c1}	I	$2\cdot10^{-7}$	0,4874	0,987
	II	$4\cdot10^{-7}$	0,5212	0,953
τ_{c2}	I	$4\cdot10^{-6}$	0,4312	0,953
	II	$2\cdot10^{-6}$	0,4196	0,983
$\dot{\gamma}_{cr}$	I	$2\cdot10^{-5}$	0,3293	0,923
	II	$3\cdot10^{-7}$	0,533	0,999

6 Inclined plane results

The inclined plane allows measurement of the critical stress separating the solid and the liquid regime closer to the field conditions (i.e., flow over steep slopes) according to equations (4) and (5) reported in section 2. This test has been performed at the same solid volumetric concentration analysed with the rheometrical tests on materials I and II. The main results are that the dynamic and static yield stress increase with the solid concentration and that the static yield stress is always higher than the dynamic one. Both results are in agreement with the rheometrical ones. The comparison between the results of both tests is reported in fig. 7a for τ_{c1} and in fig. 7b for τ_{c2}. It has been observed that the materials analyzed show good agreement between the rheometrical results and the inclined plane results. This comparison suggests that the technique of the inclined plane might be used for determining the basic rheological parameters of materials including the whole range of particle sizes.

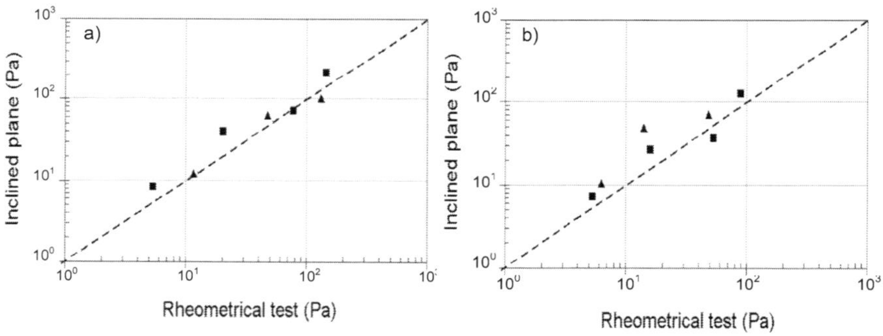

Figure 7: Comparison between the experimental results obtained from the inclined plane test and the experimental results obtained from the rheometrical test at different solid concentrations in terms of: a) static yield stress and b) dynamic yield stress. The filled triangles are relative to material I and the filled rectangles are relative to material II.

7 Conclusion

This paper concerns the study of the post-failure behaviour of pyroclastic debris flows in the Campania region (Italy). Two natural pyroclastic soils have been sampled and remixed with distilled water at different solid fractions. The behaviour of these mixtures has been investigated with a vane rotor rheometer and an incline plane. The principal result is the identification of a specific range of solid contents in which the mixtures could be considered as homogeneous fluid and could be tested with rheometrical apparatuses. Beyond these solid concentration ranges, the material mixtures behave like solids, while below them, the mixtures behave like pure liquids. For the materials analyzed the transition between "solid-like" to "fluid-like" behaviour occurs in a small range of solid concentrations and these ranges are quite different for each material, depending on the in situ porosity and the grain size distribution.

In the "fluid-like" behaviour the material mixtures behave like a yield stress fluid, and a classical Herschel and Bulkley model reproduces well the experimental data. Nevertheless, a hysteresis effect, associated with the instability of the material behaviour, has been observed for the high solid volumetric concentrations. The material starts to flow beyond a critical stress at a relatively large shear rate. These results suggest that in the field during the debris flow motion, a small variation of the solid fraction can lead to changing the behaviour from "solid-like" to "fluid-like" and vice versa. The critical shear rate related to low shear stress for lower solid fractions might explain the in situ observed post-failure behaviour of pyroclastic debris flows, which are able to flow over very long distances even over smooth slopes. Finally, the inclined plane technique could be used in the field for determining the static and dynamic yield stresses of natural pyroclastic suspensions.

References

[1] Cascini, L. & Sorbino, G., The contribution of soil suction measurements to the analysis of flowslide triggering. Proc. Int. Workshop on Occurrence and Mechanisms of Flows in Natural Slopes and Earthfills – IW-Flows2003, Sorrento, pp 77-86, 2003.

[2] Scotto di Santolo, A., Le colate rapide. Helvelius Edizioni s.r.l., 2002.

[3] Scotto di Santolo, A., Analisi geotecnica dei fenomeni franosi nelle coltri piroclastiche della provincia di Napoli. PhD thesis, University of Naples "Federico II" and Rome "La Sapienza", 2000a.

[4] Scotto di Santolo, A., Analysis of a steep slope in unsaturated pyroclastic soils. Proc. Asian Conference on Unsaturated Soils, Singapore, pp 569-574, 2000b.

[5] Rickenmann, D. & Koch, T., Comparison of debris flow modelling approaches. Proc. 1st Int. Conf. Debris flow Hazard Mitigation: Mechanics, Prediction, and Assessments, San Francisco, pp 576- 585, 1997.

[6] McDougall, S. & Hungr, O., A model for the analysis of rapid landslide motion across three-dimensional terrain. Canadian Geotechnical Journal 41, pp 1084-1097, 2004.

[7] Revellino, P., Hungr, O., Guadagno, F.M. & Evans, S.G., Velocity and runout prediction of destructive debris flows and debris avalanches in pyroclastic deposits, Campania region, Italy. Environmental Geology, 45, pp 295-311, 2004.

[8] Scotto di Santolo, A. & Evangelista, A., Some observations on the prediction of the dynamic parameters of debris flows in pyroclastic deposits in the Campania region of Italy. Int. Journal of Natural Hazards 50, pp 605-622, 2009.

[9] Scotto di Santolo, A., Pellegrino, A. M. & Evangelista, A., Experimental study on the rheological behaviour of debris flow material in Campania region, Fifth International Conference on Computational and Experimental Methods in Multiphase and Complex Flow, New Forest, pp 305-316 (2009).
[10] Papa, R., Indagine sperimentale di una copertura piroclastica di un versante della Campania, PhD thesis, University of Naples "Federico II", 2007.
[11] Van Wazer, R.J., Viscosity and flow measurement (a laboratory handbook of rheology). Interscience Publishers (New York), 1963.
[12] Coussot, P., Rheometry of Pastes, Suspensions and granular materials: Application in Industry and Environmental. A John Wiley & Sons, Inc., Publications, 2005.
[13] Ovarlez, G., Rodts, S., Chateau, X. & Coussot, P., Phenomenology and physical origin of shear localization of the shear banding in complex fluids. Rheologica Acta, 48, pp 831-844, 2009.
[14] Coussot, P., Mudflow Rheology and Dynamics, IAHR Monograph Series, A.A. Balkema: Rotterdam, 1997.

Section 2
Debris flow triggering

The triggered mechanism of typhoon-induced debris flows and landslides over mainland China

G. P. Zhang, J. Xu, F. W. Xu, L. N. Zhao, Y. M. Li, J. Li, X. D. Yang & J. Y. Di
National Meteorological Center, Chinese Meteorological Administration, China

Abstract

Typhoon-induced rainstorms can trigger debris flow and landslide, causing severe losses and casualties in China. Analysis of antecedent precipitation (P_A), threshold precipitation (P_C), mean precipitation intensity (P_M), precipitation duration (T_D) and the lag time (T_L) for typhoon- and non-typhoon-induced shows that: 1) P_C is greater and P_A is lower for typhoon-induced rainstorms. For typhoon-induced rainstorms, when P_A is within 50–100mm and P_C is greater than 200mm/d, landslides and debris flows are mostly likely to happen. As for non-typhoon-induced rainstorms, P_A is within 100–150mm and P_C is within 150–200mm/d. 2). After one day of typhoon precipitation, debris flow and landslides are more likely to happen. However, for non-typhoon-induced rainstorms it is usually 2–3 days. 3) For typhoon-induced rainstorms, 75% of debris flow and landslides happen during the day when maximum precipitation intensity occurs; for non-typhoon-induced rainstorms, 45% of hazards happen 2–12 days after the maximum precipitation day. 4) Typhoon-triggered debris flow and landslides have a lower environmental danger value compared to those that are non-typhoon triggered.
Keywords: debris flow and landslide, typhoon, non-typhoon, China.

1 Introduction

China is one of the countries that has the most landfall typhoons. In the coastal areas of south-east and southern China, due to the north-to-south direction of the

WIT Transactions on Engineering Sciences, Vol 67, © 2010 WIT Press
www.witpress.com, ISSN 1743-3533 (on-line)
doi:10.2495/DEB100061

mountain ranges, the terrain plays an important role in strengthening the precipitation intensity. When a typhoon meets the cold air from the north, the precipitation will also be intensified. The break-record precipitation is always induced by landing typhoons in southeast China. The daily precipitation of landing typhoons is about 300–900mm, with several cases more than 1000mm. The storm induced by the typhoon triggers a great number of debris flows and landslides in southeast China.

Many studies show that for typhoon- and non-typhoon-induced rainstorms, the precipitation triggering mechanism is quite different. The duration of precipitation for typhoon-triggered debris flows and landslides is less than that of non-typhoon-triggered debris flows and landslides in Zhejiang province, southeast China; the antecedent and threshold precipitation for typhoon-induced debris flows and landslides are both higher than that of non-typhoon-induced debris flows and landslides [1–3]. Typhoon-triggered debris flows often happen within one hour of the moment of peak precipitation in Chinese Taiwan province [4]. Typhoons trigger shallow landslides followed by debris flows in Hong Kong [5]. For typhoon-triggered debris flows and landslides, antecedent precipitation is not the most important factor [6].

This paper plots the typhoon zonation map and analyzes the mechanism of typhoon- and non-typhoon-triggered debris flows and landslides.

2 Mapping typhoon intensity

To analyze the typhoon- and non-typhoon-triggered debris flows and landslides, a zonal map is needed. So, within the typhoon influence area, the landslides and debris events can be classified as typhoon- and non-typhoon-triggered.

The track data of tropical cyclones provided by the Joint Typhoon Warning Center (JTWC) of the USA navy over the western North Pacific, including the South China Sea from 1950 to 2009, are spatially mapped with GIS software. The typhoon routes are plotted as a line shape. The intensity of typhoons is spatially mapped with the following formula:

$$T_{i,j} = \sum_{n=1}^{N} \sum_{m=1}^{M_n} S_m \times L \tag{1}$$

where, $T_{i,j}$ is the intensity of the typhoon at grid site (i, j), the unit is km/km^2; N is the total number of typhoon routes; M_n is the total number of components when the nth typhoon route is cut into small parts with unit length of L and S_m is the Boolean value, defined as follows:

$$S_m = \begin{Bmatrix} = 1, d_{mij} \le R \\ = 0, d_{mij} > R \end{Bmatrix} \tag{2}$$

where d_{mij} is the geographical distance from the centre of grid (i,j) to the mth partition of typhoon route and R is the average radius of the typhoon.

The total number of typhoon routes is 1888, the average radius of typhoons is supposed to be 400km. The paper uses the spatial analysis model of ArcGIS software to fulfil the above work. The raster map is generated and then converted to a contour map, see fig. 1.

From Fig. 1, the typhoon-influenced area of China can be divided into three regions. Region I, including Hainan Island and Taiwan Island, is the most serious. Region II includes Zhejiang, Fujian and Guangdong Province. Region III includes all the coastal provinces except for region I, II, and many inland provinces that are attacked frequently by typhoons, see fig. 2.

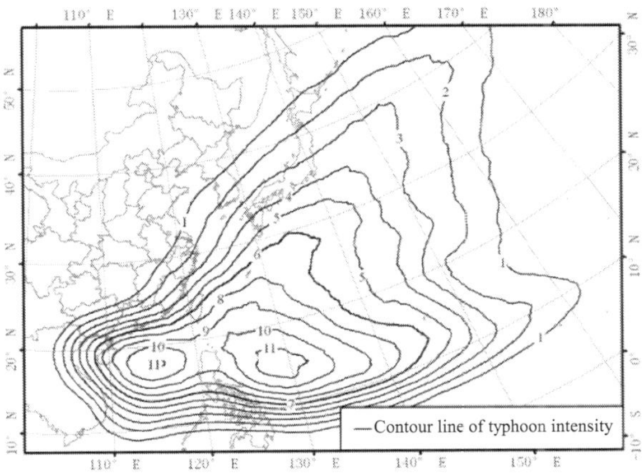

Figure 1: The spatial distribution of typhoon intensity.

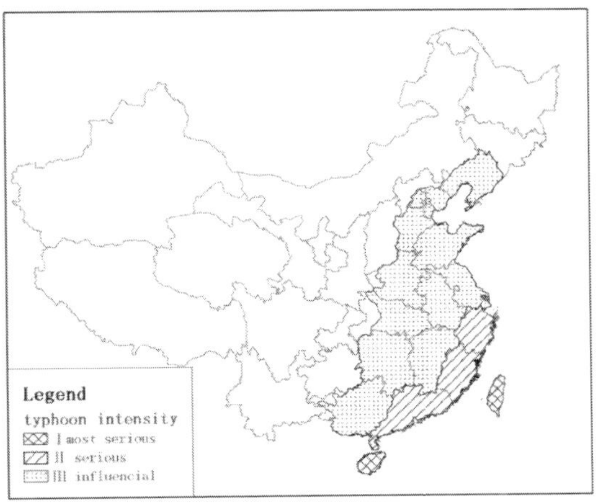

Figure 2: The zonation of typhoon intensity.

3 Precipitation analysis between typhoon- and non-typhoon-induced debris flows and landslides in regions II and III

To analyze the difference between typhoon and non-typhoon induced debris flows and landslides, eight cases of typhoon-induced storms and eight cases of non-typhoon-induced storms were selected based on the integrity of the events of debris flows and landslides, as shown in table 1. For each debris flow and landslide event, the 15 days precipitation data before the event were processed. All of the analysis is based on the hazards and precipitation data as shown in table 1.

The debris flow and landslide hazard during the period listed in table 1 is extracted from the hazard database. The precipitation observation is processed and interpolated to the hazard site.

The antecedent precipitation (P_A), threshold precipitation (P_C), mean precipitation intensity (P_M), precipitation duration (T_D) and lag time (T_L) are calculated and analyzed for typhoon-induced and non-typhoon-induced storms. The antecedent precipitation is calculated with the formula below:

$$P_A = \sum_{i=1}^{15} 0.8^i P_i \tag{3}$$

where P_A is antecedent precipitation and P_i is the ith day precipitation before the debris flow and landslide event.

Since the precise hour of the debris flow and landslide is not recorded, P_C in the paper is approximately replaced with the precipitation of the day when the debris flow and landslide happened. P_M refers to the average daily precipitation within continuous precipitation days before the landslide and debris happened. T_D refers to the precipitation duration in days before the landslide and debris flow happened. T_L refers to the number of days after the maximum precipitation day.

Table 1: The list of landslide and debris hazards for precipitation analysis.

Typhoon		Non-typhoon	
time	Hazards number	time	Hazards number
1990/6/30	10	1995/6/1-3	52
1996/8/1-8	83	1998/6/11-27	528
1999/9/4	193	1998/7/1-3	102
2001/7/1-8	38	1998/7/21-23	166
2002/8/5-13	49	2001/6/11-17	24
2002/8/19-21	10	2002/6/12-21	38
2002/9/16-18	26	2003/7/4-16	36
2004/8/13-14	103	2004/7/18-21	27
total	512	total	973

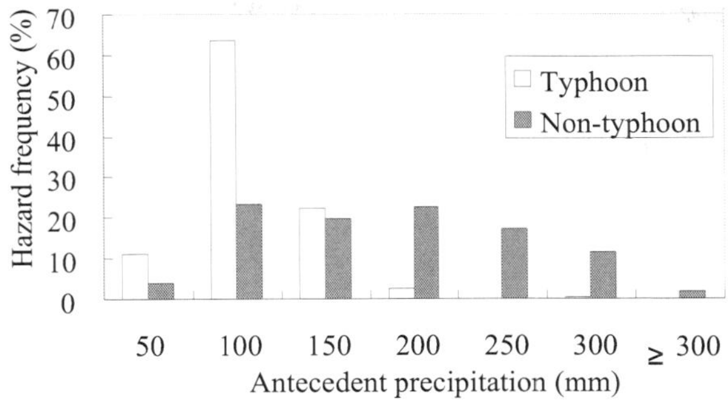

Figure 3: Antecedent precipitation (P_A) for typhoon- and non-typhoon-triggered debris flows and landslides.

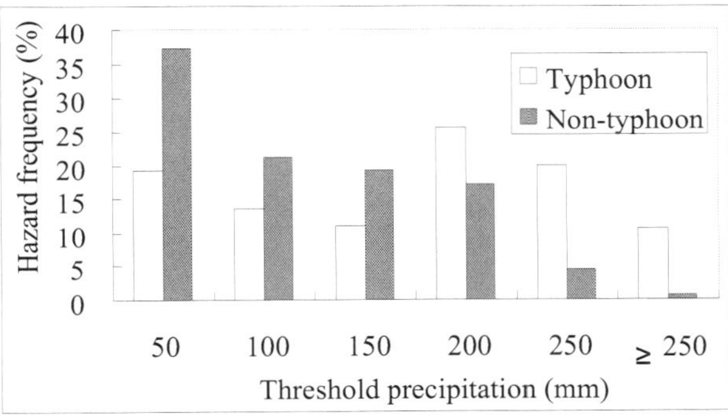

Figure 4: Threshold precipitation (P_C) for typhoon- and non-typhoon-triggered debris flows and landslides.

3.1 P_A and P_C analysis

Fig. 3 shows the P_A distribution for typhoon- and non-typhoon-induced storms. Compared to non-typhoon-, typhoon-induced debris flows and landslides need less P_A. It is mostly within 50–100mm for typhoon-induced debris flows and landslides, but for non-typhoon-induced debris flows and landslides it is mostly within 100–300mm.

Fig. 4 shows the P_C distribution for typhoon- and non-typhoon-induced debris flows and landslides. Compared to non-typhoon-, typhoon-induced debris flows and landslides need greater P_C. It is mostly within 0–50mm/d for non-typhoon-induced debris flows and landslides, but for typhoon-induced debris flows and landslides it is mostly more than 200mm/d.

The relation between P_C and P_A for typhoon-induced and non-typhoon-induced debris flows and landslides is plotted in figs. 5 and 6, respectively. P_C is decreasing while P_A is increasing both for typhoon- and non-typhoon-induced debris flows and landslides, but their features are quite different.

When P_A is within 0–100mm, greater P_C is needed to trigger typhoon-induced debris flows and landslides, and when P_A is greater than 100m, lower P_C is needed, see fig. 5. When P_A is within 50–100mm, 27% of debris flows and landslides happened when P_C is greater than 200mm. Greater P_A is needed for non-typhoon-induced debris floss and landslides, see fig. 6. Nearly 52.8% of debris flows and landslides happened when P_A was greater than 150mm.

The difference between typhoon- and non-typhoon-induced debris flows and landslides is that P_C is greater and P_A is lower for those that are typhoon induced. When P_A is within 50–100mm and P_C is greater than 200mm/d, typhoon-induced debris flows and landslides are mostly likely to happen, but non-typhoon-induced debris flows and landslides are most likely when P_A is within 100–150mm and P_C is within 150–200mm/d.

3.2 Comparison of T_D between typhoon- and non-typhoon-triggered debris flows and landslides

For typhoon-triggered debris flows and landslides, T_D is mainly within 2–3 days, in which more than 70% of hazards happened. However, for non-typhoon-

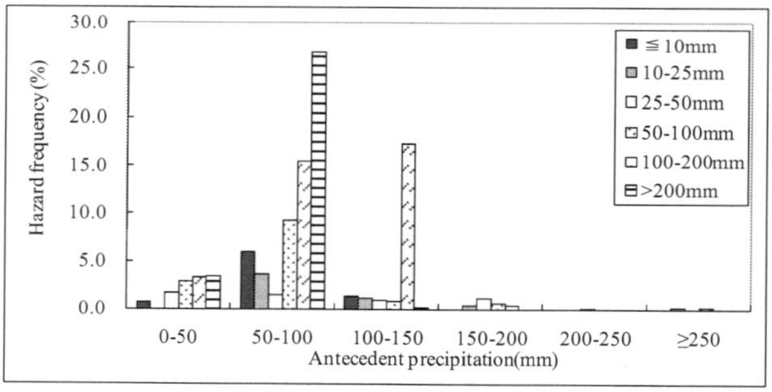

Figure 5: Antecedent precipitation (P_A) and threshold precipitation (P_C) for typhoon-induced debris flows and landslides.

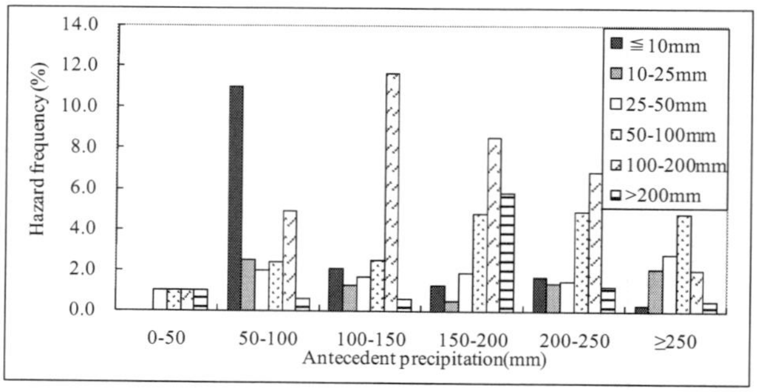

Figure 6: Antecedent precipitation (P_A) and threshold precipitation (P_C) for non-typhoon-induced debris flows and landslides.

triggered debris flows and landslides, T_D is in a greater range, and it shows a relatively high peak within 4–5 days, see fig. 7.

It can be seen from fig. 7 that after one day of typhoon precipitation, debris flows and landslides are more likely to happen. However, for non-typhoon-triggered debris flows and landslides it is usually 2–3 days.

The relation of T_D and P_M for typhoon- and non-typhoon-triggered debris flows and landslides are plotted in fig. 8. Although for typhoon- and non-typhoon-triggered debris flows and landslides the T_D are both great, they are temporally quite different. The precipitation for typhoon-triggered debris flows and landslides is mostly distributed within or one day before the hazards happened, but for non-typhoon-triggered debris flows and landslides this is 1–7 days after the typhoon.

3.3 Lag time after the maximum precipitation intensity

The period after the maximum precipitation day is called lay time (T_L) and is plotted, for typhoon- and non-typhoon-triggered debris flows and landslides, in fig. 9. It can be seen that 75% of debris flows and landslides happened during the maximum precipitation day of the typhoon, and 10% happened just one day before the maximum precipitation day.

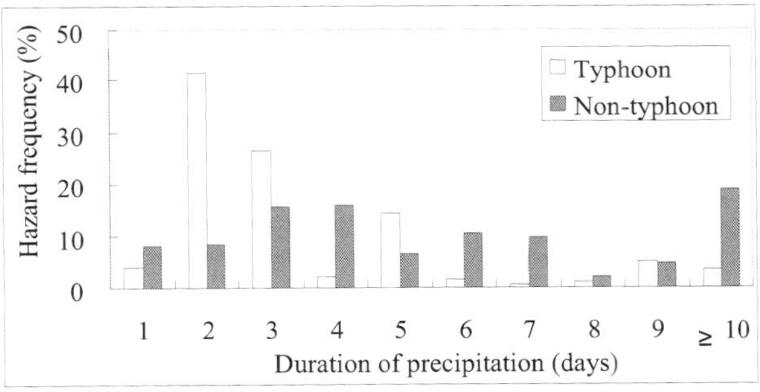

Figure 7: Duration for typhoon- and non-typhoon-triggered debris flows and landslides.

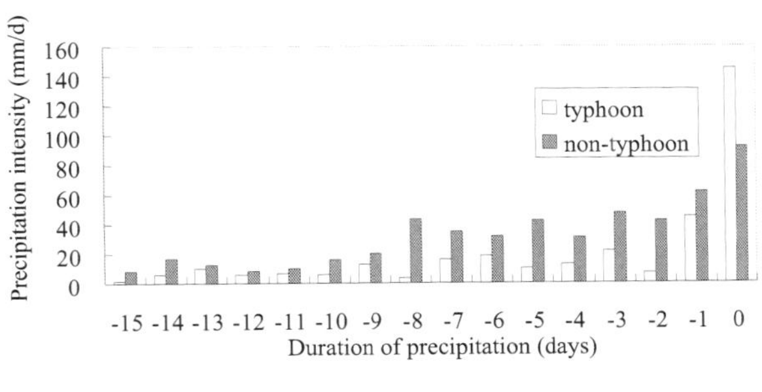

Figure 8: Duration and mean intensity of precipitation for typhoon- and non-typhoon-triggered debris flows and landslides.

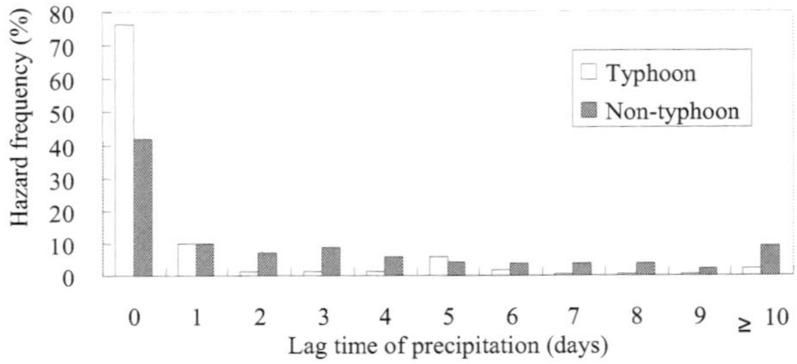

Figure 9: Lag time after the maximum precipitation day for typhoon- and non-typhoon-triggered debris flows and landslides.

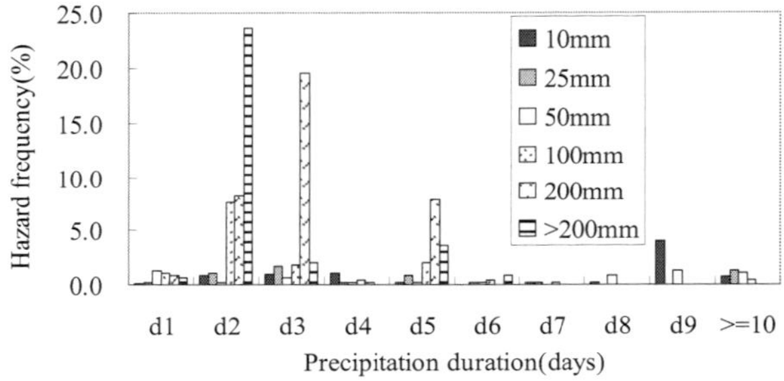

Figure 10: Precipitation duration (T_D) and threshold precipitation (P_C) for typhoon-triggered debris flows and landslides.

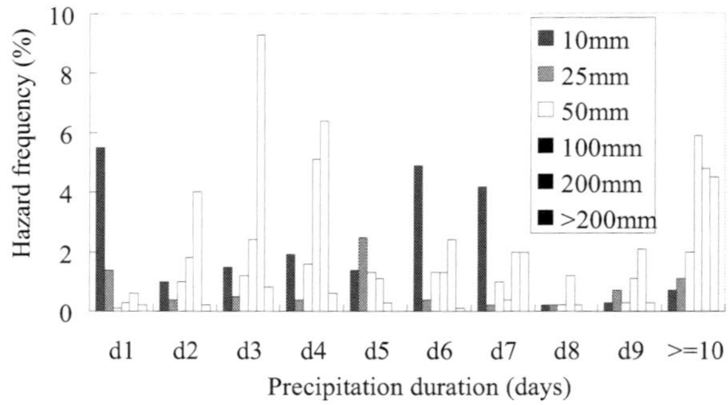

Figure 11: Precipitation duration (T_D) and threshold precipitation (P_C) for non-typhoon-triggered debris flows and landslides.

Non-typhoon-triggered debris flows and landslides are roughly the same as typhoon-triggered debris flows and landslides during or one day before the maximum precipitation day, but there still 45% of hazards happened 2–12 days after the maximum precipitation day.

3.4 Analysis of T_D and P_C

It is shown that shorter T_D corresponds to lower P_C, see fig. 10. For typhoon-triggered debris flows and landslides, 23.7% happen when T_D is two days and P_C is greater than 200mm/d; while 19.6% happen when T_D is three days and P_C is within 100-200mm/d.

For non-typhoon-triggered debris flows and landslides, the frequency is distributed more evenly, see fig. 11. When T_D is 3–4 days and P_C is within 100–200mm/d, the debris flows and landslides are more likely to happen. They are also more likely to happen when T_D is greater than 9 and P_C is greater than 50mm.

4 The environmental background in region II

The environmental background, which is made up of several topography, geology and land-use factors, plays an important role for debris flow and landslide occurrence. Six factors, altitude, aspect, slope, lithology, geological fault line density and land-use, are taken into consideration. Each factor is rasterized and reclassified with GIS tools, then the Information Model is used:

$$I_i = \sum_{j=1}^{m_i} \ln\left(\frac{N_{i,j} / N}{S_{i,j} / S}\right) \qquad I = \sum_{i=1}^{6} W_i I_i \tag{4}$$

where, m_i is the number of classes for factor X_i, N is the total number of debris flow and landslide hazard occurrences, $N_{i,j}$ is the total number of debris flows and landslides at the pixel where the factor X_i is equal to $j (j=1,2,...,m_i)$, S is the total number of pixels within the research area, $S_{i,j}$ is the total number of the pixels where factor X_i is equal to j, W_i is the weight of the factor X_i, I_i is the information for factor X_i, and I is total information.

The value of I reflects the debris flow and landslide hazard vulnerability. It is reclassified to be the hazard danger value. The environmental danger value for typhoon- and non-typhoon-triggered debris flows and landslides are extracted respectively and plotted in fig. 12. The environmental background is described with a danger value of 1–5, the larger the value, the more likely the debris flow and landslide is to happen.

The danger zoning map shows the vulnerability to debris flows and landslides. Both typhoon- and non-typhoon-triggered debris flows and landslides happen more frequently when environment danger value is increased, see fig. 12. However, the two triggers are quite different. For typhoon-triggered debris flows and landslides, a lower danger value of environmental background is needed when compared to non-typhoon-induced debris flows and landslides.

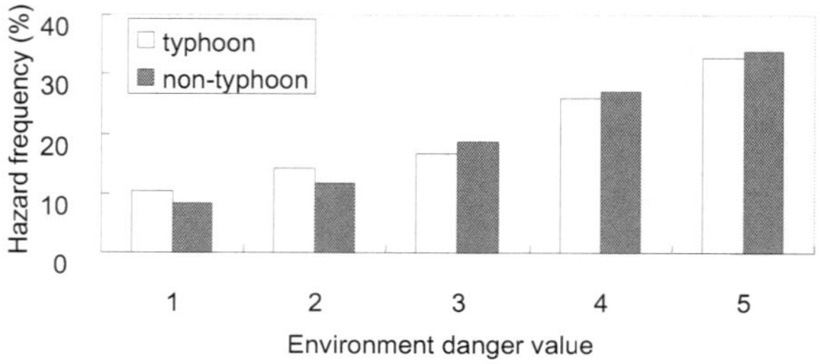

Figure 12: The occurrence of typhoon- and non-typhoon-triggered debris flows and landslides under the varied environment background.

5 Conclusions

The typhoon-influenced area of China can be divided into three regions. The typhoon intensity is has neither increased nor decreased persistently for the entire region II since 1950. It varies from north to south within different decades.

The difference between typhoon- and non-typhoon-triggered debris flows and landslides is that P_C is greater and P_A is lower for those induced by typhoons. Typhoon-induced debris flows and landslides are most likely to happen when P_A is within 50–100mm and P_C is greater than 200mm/d, but non-typhoon-triggered debris flows and landslides are most likely to happen when P_A is within 100–150mm and P_C is within 150–200mm/d. After one day of typhoon precipitation, debris flows and landslides are more likely to happen. However, for non-typhoon-triggered debris flows and landslides it is usually 2–3 days. For typhoon-triggered debris flows and landslides, 75% happened during the day when maximum precipitation intensity appeared; for non-typhoon-triggered debris flows and landslides, 45% happened 2–12 days after the maximum precipitation day.

Typhoon-induced debris flows and landslides need a lower danger value of environmental background to be triggered compared to those that are non-typhoon triggered.

Acknowledgements

This research is supported by the National Science Foundation (Grant number: 40971016), the Eleventh Five-Year Plan Project of Chinese Meteorological Administration on meteorological monitoring and hazard early warning and the Research Fund for Commonweal Trades (Meteorology) (Grant number: GYHY200706037).

References

[1] Xie, P., Yatagai, A., Chen, M., et al., A gauge-based analysis of daily precipitation over East Asia. J. Hydrometeorology, 8(6), pp. 607–626, 2007.

[2] Du, H.L., Niu, X.X., Yin, K.L., Xie, J.M., et al., Meteorological condition analysis and forecast research of landslides and debris flows in Zhejiang province. Journal of Tropical Meteorology (in Chinese), 21(6), pp. 642–650, 2005.
[3] Yin, K.L., Zhang, G.R., Gong, R.X., et al., A real time warning system design of geo-hazards supported by Web-GIS in Zhejiang Province, China. Hydrogeology and Engineering Geology (in Chinese), 30(3), pp. 19–23, 2003.
[4] Chen, C.Y., Chen, T.C., Yu, F.C., et al., Rainfall duration and debris-flow initiated studies for real-time monitoring. Environmental Geology, 47, pp. 715–724, 2005.
[5] Dai F.C. & Lee C.F., Analysis of rainstorm-induced slide-debris flows on natural terrain of Lantau Island, Hong Kong. Engineering Geology, 51, pp. 279–290, 1999.
[6] Brand E.W., Slope instability in tropical areas. Proc. Of the 6th Int. Conf. On Landslides, eds. D. H. Bell, Rotterdam: Balkema, pp. 2031–2051, 1995.

Debris flow occurrences in Rio dos Cedros, Southern Brazil: meteorological and geomorphic aspects

M. Kobiyama, R. F. Goerl, G. P. Corrêa & G. P. Michel
Laboratory of Hydrology, Federal University of Santa Catarina, Brazil

Abstract

In Santa Catarina State (Brazil), Rio dos Cedros is one of the cities that suffered from natural disasters triggered by very intense rainfall in November 2008. According to the Municipal Civil Defense, this event caused economic losses of US$2.1 million and 96 homeless, directly affecting 90% of the municipal population. The principal phenomenon responsible for these disasters was the debris flow. In this context, field surveys were conducted throughout the city in order to investigate the possibility for houses to be affected by landslides and debris flows. After the preliminary evaluation, two cases of debris flow whose extensions were over 1 km were chosen for a more detailed survey. The objective of the present study was to analyze the meteorological and geomorphic aspects of these cases. The total rainfall in Rio dos Cedros in November 2008 and the entire year of 2008 were 644 mm and 2509 mm, respectively. An analysis of the daily rainfall during October and November 2008 and the debris flow occurrence time indicates that the factor triggering debris flows in Rio dos Cedros in 2008 was the accumulated rainfall, not its intensity. Two debris flows analyzed in the present study had different geomorphic conditions, one being confined in the channel and showing ordinal debris flow (Debris A), and the other unconfined, presenting the debris avalanche feature (Debris B). The larger density of clastic blocks was observed at the depositional area of Debris B. Both the localities are characterized with migmatite. Topographic analysis showed that the volume of the mass movement is related with topographic parameters (elevation difference and travel distance). This relationship implies that the two cases of Rio dos Cedros had similar behavior to other cases reported in temperate and cold regions. It is, therefore, concluded that there is geomorphic control on

WIT Transactions on Engineering Sciences, Vol 67, © 2010 WIT Press
www.witpress.com, ISSN 1743-3533 (on-line)
doi:10.2495/DEB100071

the debris flow behavior. The similarity of Debris A and B implies the high potential effects of woody vegetation on the debris flow feature.

Keywords: debris flow, volume of mass movement, topographic survey, rainfall, woody vegetation, Brazil.

1 Introduction

By elaborating an atlas of natural disasters of Santa Catarina State (SC), Brazil, Herrmann [6] shows that SC has frequently suffered from hydrological disasters. In November 2008, the extremely intense rainfall event triggered floods and landslides in SC, especially in the Itajaí Valley. This event might be the worst in the whole history of SC. Rocha *et al.* [9] analyzed the daily rainfall data obtained in Blumenau city, which is located in the Itajaí Valley, and concluded that the daily rainfall of 347.2 mm registered on 23rd November corresponds to a return period of more than 10,000 years for this city. Among 293 cities in SC, 63 declared a state of emergency and 14 a state of public calamity in November 2008. According to the State Civil Defense Report 31/Dec/2008, which presented the total damages caused by these disasters, there were 32,853 homeless, 135 dead and 2 missing in SC.

In the hydrological and socio-economic aspects, the Itajaí Valley is one of the most important regions in SC and consists of 53 cities. According to Fraga [3] and Frank and Pinheiro [4], the floods in this valley have been registered for more than 150 years. Up to now, the historically large floods in the Itajaí Valley occurred in 1855, 1880, 1911, 1927, 1957, 1983, 1984, 1992 and 2008.

Rio dos Cedros, city located in the Itajaí Valley, at a distance from Blumenau about 30 km, declared a state of pubic calamity due to the intense rainfall in November 2008. It was reported that 8,561 peoples were directly affected, 96 homeless, no dead, and economic losses of agriculture, livestock, industry and basic sanitation infra-structures in the range of US$1.34 million, 300 thousands, 40 thousands, and 390 thousands, respectively. The floods occurred in the urban area and the typical landslide type that occurred in many rural ones was the debris flow in this city.

Though the frequency of its occurrence has increased recently, there are only a few studies on the debris flow in Brazil until now. The debris flow research can be, therefore, considered as one of the priorities in the Brazilian society. In this context, the objective of the present study was to analyze the meteorological and geomorphic aspects of two debris flow features which occurred in Rio dos Cedros city in November 2008.

2 Methods

2.1 Study area

In Rio dos Cedros city, the population and the area are 9,685 and 556 km^2, respectively. Most of the inhabitants live in the urban area (18 km^2) located on

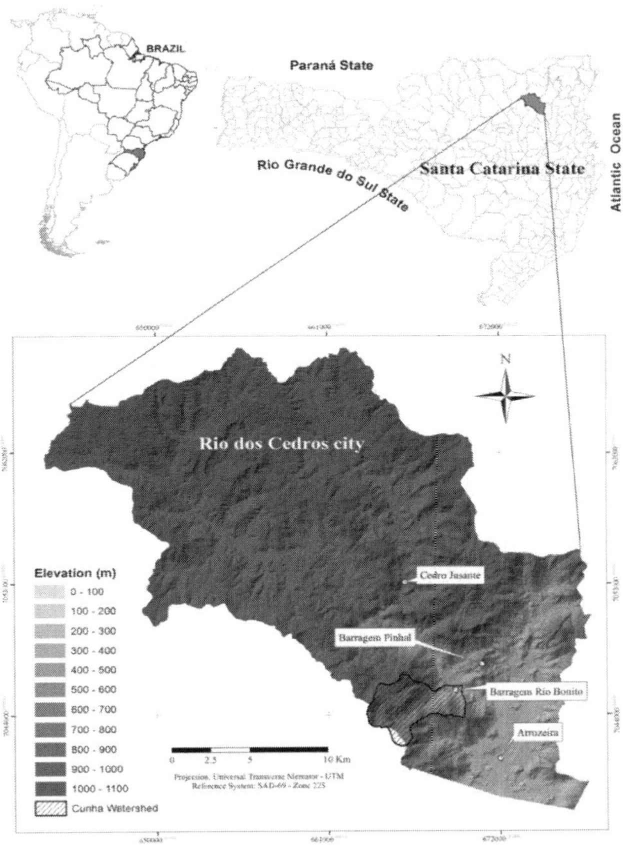

Figure 1: Localities of Rio dos Cedros city, Cunha watershed and rain gauge stations.

the floodplain. Thus, the floods have frequently affected human lives. The highland is used for agriculture, reforestation and hydro-energy generation. There are two dams (Palmeiras and Pinhal) in this city. As the administration boundary of the city coincides almost with the Cedro River watershed divide, the watershed management may be highly useful for the city hall (Figure 1). There are topographic maps only at a 1/50.000 scale for the study area.

The relief of Rio dos Cedros city can be divided into three zones: (i) the law lands formed by floodplain and by small hills with altitude from 75 to 150 m and gentle slopes (~12°); (ii) the highlands with altitude from 600 to 1100 m and gentle slopes in most of the areas (~16°); and (iii) the transition zone between the two former lands, with steep slopes and embedded valleys. In this transition zone the major landslides occurred. Goerl *et al.* [5] reported 15 severe landslides triggered by the intense rainfall in this city in November 2008. Their main type was the debris flow, and most of them occurred on steep slopes with colluvium deposits on their base. From this evidence the authors [5] concluded that landslides are very frequent and natural phenomena in the geological and geomorphologic time scale. The fact that there are a lot of big blocks along the main channels in the watershed, especially in the zones (ii) and (iii), implies the high susceptibility to the debris flows in these two zones.

2.2 Meteorological data

To analyze the rainfall characteristics in Rio dos Cedros, the present study used the monthly data obtained at the rain gauge station (Arrozeira) of the National Agency of Water – ANA from 1942 to 2006 and the hourly data from 2007 to 2008 recorded at three rain gauge stations (Barragem Pinhal, Barragem Rio Bonito, Cedro Jusante) of the Company of Agricultural Research and Rural Extension of Santa Catarina – EPAGRI. Their localities are shown in Figure 1.

2.3 Field survey

In the Cunha River watershed, two comparatively-larger debris flows were investigated. Figure 2 shows the digital elevation model of this watershed and two debris flows which are here called Debris A and Debris B. It also presents Debris A (1 and 2) and Debris B (1 and 2) in detail.

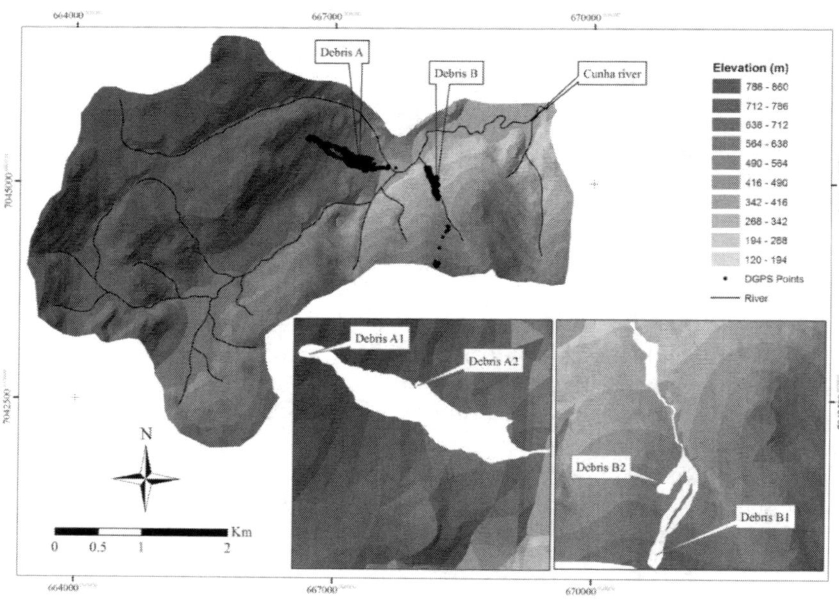

Figure 2: Localities of two debris flows (Debris A and Debris B) in the Cunha river watershed.

The field survey with the Leica Total Station and the Trimble Differential GPS (DGPS) was carried out during September 2009, about 10 months after the occurrence, because it was necessary to wait for the complete stabilization of the hillslopes. The DGPS was used to collect several points around and in the middle of the debris flow for determining the geometric relationship and for calculating the mass movement volume and the depositional volume. When it was not possible to collect data with the DGPS, the Total Station was used.

All the points were put into the same projection and reference system: UTM Zone 22S and SAD-69. More than 3000 points were collected with an altimetric error less than 0.5 m. For each debris flow, the survey focused on three parts: (i) starting; (ii) channel transporting; and (iii) depositional area.

2.4 Debris flow volume analysis

Based on the field observations, the rupture surface of the landslide starting areas was considered as an ellipsoid. This geometric assumption easily permitted to calculate the volume of mass movement. By processing the DGPS point in the ArcGIS software, the geometric relationships were established for each debris flow. The starting area, the spread channel area and the depositional area were delimitated with the reach angle and the field observations. Both the cases (Debris A and Debris B) had two different staring areas, i.e., two initial movements. In other words, four landslides occurred and formed two large debris flows. For a better presentation of the real forms, each landslide was calculated individually. Then, the Debris A (or Debris B) was divided into Debris A1 and A2 (or Debris B1 and B2).

For each starting area, the altimetry difference between the higher and lower point (*Hs*) and the axial distances (l_s) were calculated, and then $\tan\theta$ (= *Hs*/*ls*) was determined. The rupture area of the movement follows an ellipse plane, passing through the hypotenuse vertex. Thus, the mass movement height (*hsi*) is considered equal to half of the median, and the hypotenuse is the length of the Y plane. Figure 3 illustrates the definition of the geometry used in this calculation.

Then, the total volume of the mass movement for each debris flow can be expressed by the following eqn:

$$V_{Debris} = V_{debris1} + V_{debris2} = \frac{\frac{4}{3}\pi a_1 b_1 c_1}{2} + \frac{\frac{4}{3}\pi a_2 b_2 c_2}{2} \tag{1}$$

where V_{Debris} is the total volume of the mass movement of the Debris A or B; $V_{debris1}$ and $V_{debris2}$ are the volume of the two initial movements; a_1 and a_2 are the positive numbers which represent the X-dimension of the two ellipsoids; b_1 and b_2 are the positive numbers which represent the Y-dimension of the two ellipsoids; c_1 and c_2 are the positive numbers which represent the Z-dimension of the two ellipsoids.

Corominas [2] and Rickenmann [7] separately proposed an equation that relates the volume of movement mass (V) with the angle of reach (β) in which $\tan\beta = H/L$:

$$V_C = \left(\frac{L}{1.03H}\right)^{\frac{1}{0.105}} \quad (2)$$

$$V_R = \left(\frac{L}{1.90H^{0.83}}\right)^{\frac{1}{0.160}} \quad (3)$$

where V_C and V_R are the volumes in m^3 estimated by Corominas [2] and Rickenmann [7], respectively; H is the elevation difference between starting point and the lowest point of the depositional area in meter; and L is the travel distance in meter. Corominas [2] analyzed the topography of 71 debris flows reported in several countries and made a linear regression. Rickenmann [7] elaborated the same relation by using the data of 140 debris flows and 51 rock avalanches recorded in Swiss Alps.

Analogically, the total volumes of the mass movement calculated by Corominas [2] and Rickenmann [7] equations can be expressed:

$$V_C = \left(\frac{L_1}{1.03H_1}\right)^{\frac{1}{0.105}} + \left(\frac{L_2}{1.03H_2}\right)^{\frac{1}{0.105}} \quad (4)$$

$$V_R = \left(\frac{L_1}{1.90H_1^{\ 0.83}}\right)^{\frac{1}{0.160}} + \left(\frac{L_2}{1.90H_2^{\ 0.83}}\right)^{\frac{1}{0.160}} \quad (5)$$

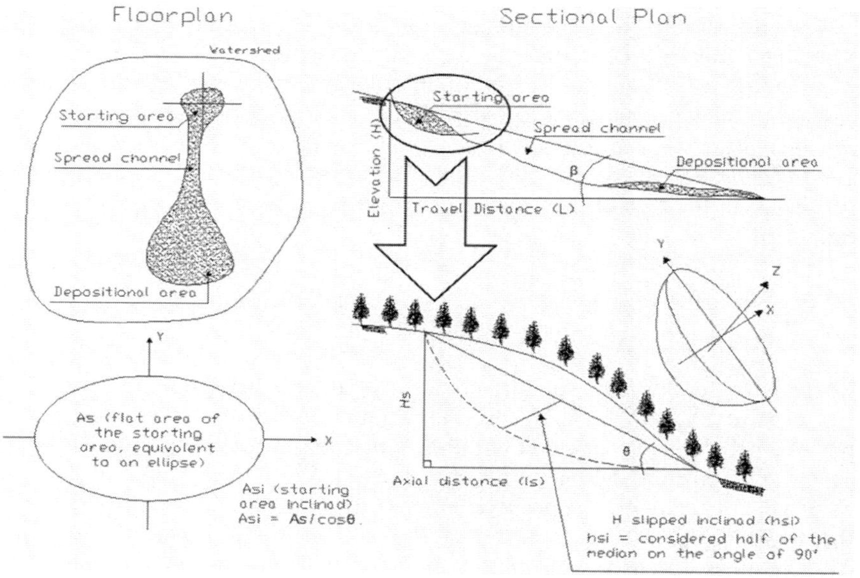

Figure 3: Geometry of debris flow.

The volume estimated with eqn (1) was compared to those obtained with eqns (4) and (5).

3 Results and discussions

3.1 Rainfall analysis

Figure 4 shows the historical data of the annual rainfall from 1942 to 2008. It can be observed a slight increase of the annual rainfall during the studied period, with its mean value of 1651 mm/year. The highest value was 2863 mm/year in 1983; meanwhile 2008 was the second rainiest year with 2509 mm/year.

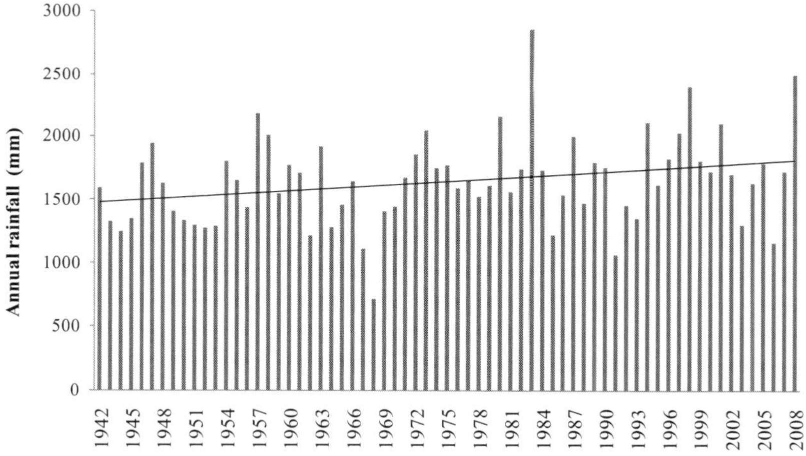

Figure 4: Annual rainfall in Rio dos Cedros during the period 1942 to 2008.

The total rainfall of October and November 2008 was 1085 mm which represented about 43.3% of the 2008 annual rainfall. On 3rd October, the most intense rainfall (122 mm/day) was recorded, and after this event no high rainfall intensity took place. In November, the values of the daily rainfall were not very high (less than 80 mm/day), but their frequency was much higher than in October and the accumulated rainfall was very high (Figure 5).

According to Mr. Rui Mayer (local resident), the mass movement took place on three stages, the first at 1:15 a.m., the second at 2:00 a.m. and the third at 3:00 on 24th November. The observed rainfall data in Rio dos Cedros showed that during the period from 0:00 a.m. to 3:00 a.m. on the same day, it rained 10 mm which might not be able to cause such a terrible and intense phenomenon. The accumulated rainfall from 0:00 a.m. on 23rd November to 3:00 a.m. on 24th November was 95 mm. Only for one week from 18th to 24th November, it rained totally 256 mm, more than twice value of the mean monthly rainfall for November. In October and November 2008, the highest values of monthly rainfall during the period 1942 to 2008 were recorded with 441 mm and 644 mm,

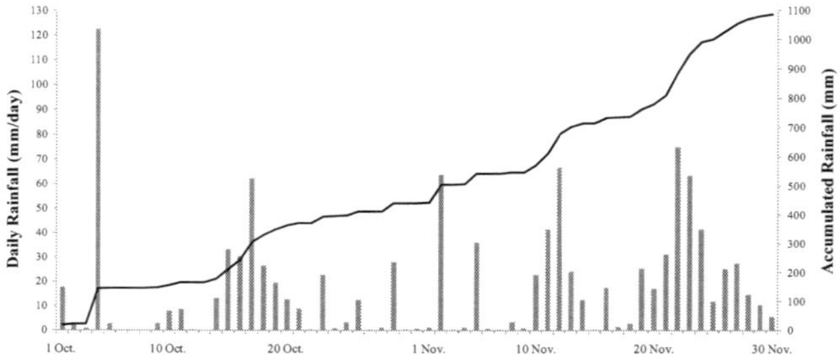

Figure 5: Daily and accumulated rainfalls in Rio dos Cedros from October to November 2008.

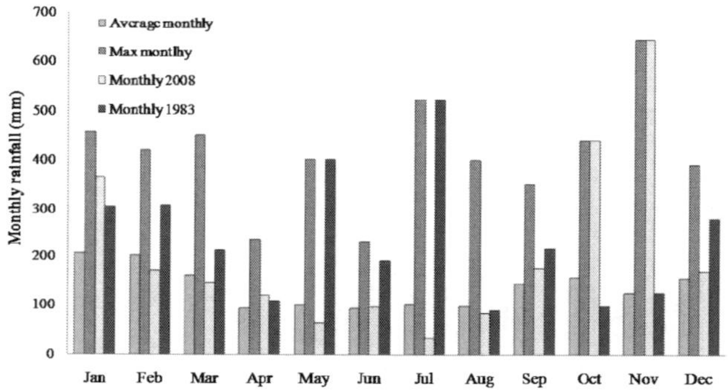

Figure 6: Monthly rainfall data from Rio dos Cedros: mean monthly rainfall from 1942 to 2007, maximum monthly record between 1942 and 2008, monthly rainfall of 2008, and monthly rainfall of 1983.

respectively (Figure 6). The monthly rainfall in November 2008 was highest during the period 1942-2008. Therefore it may be said that in this case the triggering factor was the accumulated value of the rainfall, not its intensity.

Mr. Genésio Zoboli (local resident) informed that the water of the stream where debris flow (Debris B) passed had already possessed a high turbidity since September 2008 and that in 1974 a debris flow with smaller scale and intensity occurred at the same place. It indicates that the locality of the Debris B has the high vulnerability to the debris flow.

3.2 Geomorphic analysis

From the initial movement of higher altitude, the total distance of Debris A was 1050 m, while Debris B reached 1270 m. Both debris flows occurred at the same rock type (migmatite). Table 1 shows the calculation results with eqns (1), (4) and (5) that present the geometric analysis, Corominas [2] and Rickenmann [7] relationships, respectively. It can be noted that, in both the cases A and B, the equation of Corominas [2] presented the highest value, while Rickenmann [7]

Table 1: Calculation results of different methods.

		Calculated eqn (1)	Corominas eqn (4)	Rickenmann eqn (5)	Weighted	tanθ	tanβ
		Volume (m³)					
Debris A	DebrisA1	49,794.77	19,952.92	7,458.11	89.05	0.34	-
	DebrisA2	6,120.69	935,236.77	35,521.16	10.95	0.23	-
	Total	55,915.46	955,189.69	42,979.27	100.00	-	0.33
	Difference*	-	899,274.23	12,936.19	-	-	-
	Difference (%)	-	1,608.27	23.14	-	-	-
Debris B	DebrisB1	26,135.71	12,702.70	7,071.48	42.61	0.36	-
	DebrisB2	35,201.03	62,534.86	12,002.17	57.39	0.30	-
	Total	61,336.74	75,237.56	19,073.65	100.00	-	0.33
	Difference*	-	13,900.82	42,263.09	-	-	-
	Difference (%)	-	22.66	68.90	-	-	-

*Difference between eqn (1) and eqn (4) (or eqn (5)).

the lowest. For the Debris A, the value obtained with the equation of Corominas [2] (eqn (4)) was closer to the value obtained with the topography survey, while in Debris B Rickenmann [7] (eqn (5)) was closer.

Since each debris flow had two starting areas (initial movements) (Figure 2), their weights were calculated. In the Debris A, the movement A1 had the more importance for the total volume and consequently the more contribution to the total travel distance of the debris flow. In the Debris B, both the initial movements had similar weights and volumes, contributing equally to the magnitude of the debris flow occurrence.

Figure 7 plots the values of the Debris A and B obtained with eqn (1) on the diagram originally elaborated by Rickenmann [8] which relates tanβ with the mass movement volume. In the tropical environments, the soil layer tends to become larger than that in the temperate regions. In both the cases of the present study, the average of the soil layer depth was 15 m. And the soil texture was characterized with large quantity of silt and clay. Furthermore there was a lot of vegetation on the occurrence localities, especially trunks with 20-30 m height, which might characterize the woody debris flows. In spite of these conditions, it is observed that two cases of Rio dos Cedros had the similar behavior to other cases shown by Rickenmann [8], especially to Swiss debris flows.

Though Figure 7 shows that the Debris A and B had a similar behavior, the geomorphic conditions are different between the Debris A and B. The Debris A is characterized as a debris avalanche, not channeled or constrained, meanwhile the Debris B is mostly confined and received a strong topographic control.

Rickenmann [8] mentioned that the angle of initial depositional process varies between 6 and 12° for many (larger) debris flows and can be larger than 27° for smaller and unconfined cases. Slaymaker [10] and Bryant [1] assumed that the depositional angle is smaller than 12°. The depositional angle values of the Debris A and B were 13° and 14°, respectively. The similar characteristics between the Debris A and B were the large depositional-angle, the starting slope, the depth (approximately 8-10 m) of deposition layer. The factor that controls the

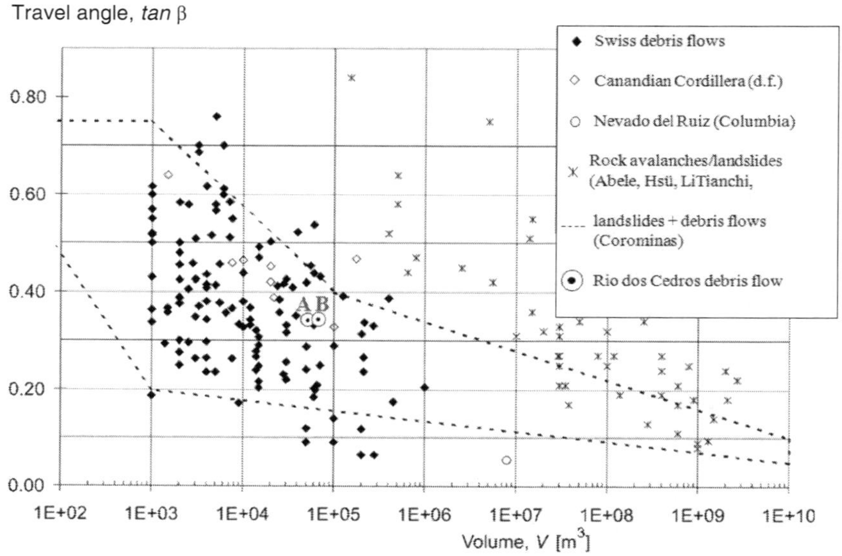

Figure 7: Travel angle vs. volume of mass movement including Rio dos Cedros debris flows A and B (modified from Rickenmann [8]).

movement and generates the similar behavior may be the high density of the woody vegetation lying on the occurrence localities. In the Debris B, the vegetation provided the dams (barriers) formation, reducing the energy gradient. But when the dam break takes place, the potential energy might be equal to the total energy and the destructive power of the debris flow was elevated. According to the residents' testimony, the movement occurrence with various waves was observed and bursts were heard.

Thus, the relationship between the mass movement volume and the travel angle (geomorphic factor) obtained in the present study is coherent to those observed in other countries. Therefore, it can be said that the geomorphic control of the debris flow is very important. But the similarity between the two debris flows which had different geomorphic conditions implies that the woody vegetation effect on the debris flow is potentially very significant.

4 Conclusions

The extremely-intense rainfall triggered the hydrological disasters in Rio dos Cedros city in November 2008. The typical disaster type was the debris flow. The present study analyzed the rainfall characteristics in this city for the period 1942 to 2008, with the special attention to the rainfall events of 2008. In October and November 2008, the monthly rainfalls were 441 mm and 643.6 mm, respectively. An analysis of the daily rainfall during these two months and the debris flow occurrence time indicates that the triggering factor in the case of debris flows in 2008 was the accumulated rainfall, not its intensity.

Based on the previous investigation, two large debris flow cases (Debris A and Debris B) were selected for the geomorphic analysis. With the field survey data of topography and the equations of Corominas [2] and Rickenmann [7], the volume of mass movement was related to the angle of reach for each debris flow. A comparison between the Rio dos Cedros' cases and data obtained in other countries permits to conclude that the geomorphic aspects of the debris flows in Rio dos Cedros are similar to those of temperate regions.

Since there are only a few studies on debris flow hazard whose disasters in Brazil have been increasing in frequency and intensity, the present study could contribute to regional prevention measures of these disasters. The debris flows research must be more enhanced in this country, by analyzing various aspects of this hydrological hazard. In the future study, the geotechnical analysis and the vegetation analysis will have to be carried out.

Acknowledgements

The present work was supported in part by the National Research Council of Brazil (CNPq) through the Grant No. 479532/2009-5. The authors are thankful to the members of the LabHidro-UFSC for support of field survey.

References

[1] Bryant, E., *Natural Hazards.* Cambridge University Press: Cambridge, 2005.

[2] Corominas, J., The angle of reach as a mobility index for small and large landslides. *Canadian Geotechnical Journal*, **33(2)**, pp. 260-271, 1996.

[3] Fraga, N.C., As enchentes no Vale do Itajaí-Açu/SC: Das obras de contenção à indústria da enchente – A problemática ambiental e a relação homem/natureza na busca de soluções. *Ra'EGA*, **5**, pp. 125-148, 2001.

[4] Frank, B. & Pinheiro, A., (eds). *Enchentes na Bacia Itajaí: 20 anos de experiências.* Edifurb: Blumenau, 2003.

[5] Goerl, R.F., Kobiyama, M., Lourenço, L.L. & Grando, A., Características gerais dos escorregamentos ocorridos em novembro de 2008 nos municípios de Bruscque, Rio dos Cedros e Timbó – SC. *Proc. of the XIII Simpósio Brasileiro de Geografia Física Aplicada*, UFV: Viçosa, CD-rom 16 pp., 2009.

[6] Herrmann, M.L.P. (org). *Atlas de Desastres Naturais do Estado de Santa Catarina.* IOESC: Florianópolis, 2007.

[7] Rickenmann, D., Empirical relationships for debris flows. *Natural Hazards,* **19(1)**, pp. 47-77, 1999.

[8] Rickenmann, D., Runout Prediction Methods. *Debris-flow Hazards and Related Phenomena,* ed. M. Jakob & O. Hungr, Springer: Berlin, pp. 305-324, 2005.

[9] Rocha, H.L., Kobiyama, M. & Silva, C.G., Análise estatística de chuvas intensas ocorridas nos municípios de Blumenau e Rio dos Cedros, SC, no

período de agosto de 2008 a janeiro de 2009. *Proc. of the XVIII Simpósio Brasileiro de Recursos Hídricos*, ABRH: Campo Grande, CD-rom 14pp., 2009.

[10] Slaymaker, O., The distinctive attributes of debris torrents. *Hydrological Sciences Journal*, **33(6)**, pp. 567-573, 1988.

Soil moisture retrieval with remote sensing images for debris flow forecast in humid regions

Y. Zhao[1,2,3], H. Yang[1,2] & F. Wei[1,2]
[1]*Key Laboratory of Mountain Hazards and Surface Process, Chinese Academy of Sciences, China*
[2]*Institute of Mountain Hazards and Environment, Chinese Academy of Sciences, China*
[3]*Graduated University, Chinese Academy of Sciences, China*

Abstract

Soil moisture is a key parameter in debris flow prediction for its influence on the critical rainfall triggering debris flow. Soil moisture can be obtained by ground measurement. However, it is difficult to extend these limited observing data to the regional scale because of the heterogeneity of land surface. The Temperature-Vegetation Dryness Index (TVDI) is a common method of estimating regional soil moisture by the images of MODIS, because of its moderate spatial resolution and high temporal resolution. However, because the basic assumption of the TVDI method is that pixels from the study region can cover the entire range of soil moisture conditions and vegetation fractions, it is difficult to determine the actual dry edge of the space in humid regions. The Crop Water Shortage Index (CWSI) calculated by actual evapotranspiration and potential evapotranspiration does not need fitting to the dry edge and wet edge. CWSI further considers about the influence of vegetation. In this paper, we applied both TVDI and CWSI methods to retrieving soil moisture using remote sensing and meteorological data in Zhejiang Province, which has a humid climate. Among CWSI, the actual and potential evapotranspiration are calculated by the SEBS model. CWSI can also directly express the extent of soil moisture. In surface soil (0-10cm), the correlation coefficient of CWSI and measured relative soil moisture (RSM) reached -0.89.
Keywords: remote sensing, soil moisture, TVDI, CWSI, debris flow forecast, humid region.

WIT Transactions on Engineering Sciences, Vol 67, © 2010 WIT Press
www.witpress.com, ISSN 1743-3533 (on-line)
doi:10.2495/DEB100081

1 Introduction

Soil moisture can affect the shear strength and infiltration capacity of soil, which can further influence the critical rainfall triggering debris flow. Therefore, soil moisture monitoring has important significance for debris flow forecasting. Due to lack of soil moisture observation stations, antecedent rainfall is used in debris flow forecast instead at present [1, 2]. Along with the development of remote sensing, the method of retrieving regional soil moisture for debris flow prediction with images of remote sensing is becoming available.

There are two indices for soil moisture estimation, Temperature-Vegetation Dryness Index (TVDI) [3] and Crop Water Shortage Index (CWSI) [4, 5]. Many researches have shown that land surface temperature is negatively related to vegetation cover fraction, and the relationship is affected by soil moisture [6-8]. TVDI is computed through the feature space constituted by vegetation index and surface temperature. It has been widely used for its simple algorithm [9-11] without meteorological data. If the study area is large enough to cover land surfaces with the whole range of soil moisture conditions and vegetation densities, “dry edge” and “wet edge” of the space can be determined by data fitting. The precondition cannot be fully satisfied in humid regions, so the fitted edges are experiential edges. Land surface actual evapotranspiration (*LE*) correlates with soil moisture. It can be estimated by remote sensing data (include vegetation index, surface temperature and albedo) united with meteorological data. So CWSI, which is calculated by *LE* and potential evapotranspiration (LE_{wet}), is also related with soil moisture. The Surface Energy Balance System (SEBS) [12] is one of estimating *LE* model according to the surface energy balance equation. SEBS is first built to estimate atmospheric turbulent fluxes and evaporative fraction using remote sensing and meteorological data. It gives a physical description to the key parameter in surface energy flux estimation-the roughness length for heat transfer, which increases calculation accuracy of *LE*. It has been used in some regions of China [13, 14].This paper will discuss how to use the images of MODIS (Moderate Resolution Imaging Spectroradiometer) and meteorological data to estimate soil moisture. CWSI calculated by SEBS and TVDI are applied to retrieve soil moisture respectively. The results of both retrieving methods are tested by ground soil moisture observations in Zhejiang Province, China.

2 Methodology

2.1 Method of TVDI [3, 6, 7, 15]

Vegetation index provides useful information about vegetation growth status while soil moisture conditions can be reflected by surface temperatures. These provide the basis for regional soil moisture monitoring with remote sensing data. When vegetation is water-stressed, the feature space composed by surface temperature (T_s) and normalized difference vegetation index (*NDVI*) is trapezoidal, otherwise it is triangular. The triangular space can be considered as a special case of trapezoidal spaces, so in this paper we used the trapezoidal space (see Fig. 1). Line 2-4 is dry edge of the space. It defines the relationship between T_s and *NDVI* in drought conditions. Line 1-3 is the wet edge and defines the relationship in wet conditions. For a given point inside the trapezoid, *TVDI* is defined as

$$TVDI = \frac{T_s - T_{s\min}}{T_{s\max} - T_{s\min}} \tag{1}$$

where T_{smin} and T_{smax} are temperature values on the wet edge and dry edge corresponding to the point, respectively. Dry edge and wet edge can be obtained by following fitting equations:

$$T_{s\max} = a + b * NDVI \tag{2}$$

$$T_{s\min} = c + d * NDVI \tag{3}$$

where a, b, c, d are fitting coefficients.

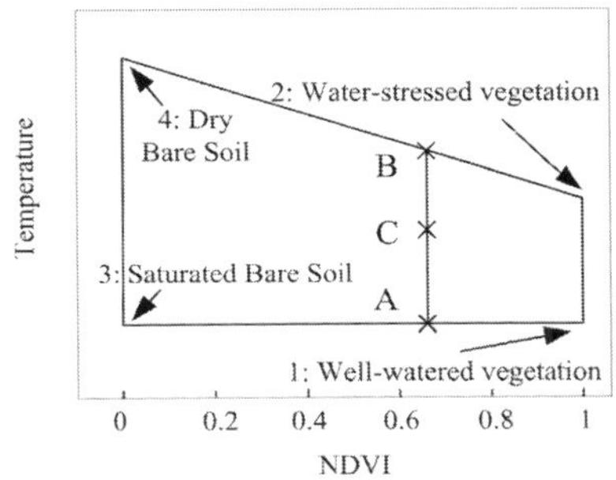

Figure 1: The hypothetical trapezoidal shape that would result from the relation between temperature and NDVI. This sketch map is modified from Moran et al. [15].

2.2 Method of CWSI [4, 5]

CWSI was originally proposed on the basis of the energy balance, and its initial prototype is the ratio of canopy-air temperature difference and air saturation deficit. In this paper, the higher the soil moisture content is, the closer the actual latent heat flux is to LE_{wet}. Otherwise the actual latent heat flux is closer to 0. So we can define CWSI to characterize the soil drought degree:

$$CWSI = 1 - LE/LE_{wet} \tag{4}$$

Different from the previous CWSI calculation method, *LE* is calculated by SEBS model. SEBS[12] is a single-source model based on the energy balance, soil heat flux *G*, net radiation R_n and sensible heat flux *H* are calculated with *NDVI*, T_s, geodata and meteorological data, and then latent heat flux *LE* is computed as:

$$LE = R_n - G - H \tag{5}$$

The Penman-Monteith formula with removing the surface resistance term is used to calculate latent heat flux in wet conditions:

$$LE_{wet} = \frac{\Delta(R_n - G) + \rho C_p (e_*(T_a) - e_a)/r_{ah}}{\Delta + r} \tag{6}$$

$$e_*(T_a) = 0.6108 \exp\left(\frac{17.27T_a}{237.3 + T_a}\right) \tag{7}$$

$$\Delta = \frac{4098 e_*(T_a)}{(237.3 + T_a)^2} \tag{8}$$

where ρ is air density, kg/m^3, C_P is specific heat capacity at constant pressure of the air, J/(kg·K), $e_*(T_a)$ is saturation vapour pressure corresponding to T_a, kPa, e_a is actual vapour pressure, kPa, r_{ah} is the aerodynamic resistance for heat transfer, s/m, Δ is the slope of vapour pressure-temperature curve, kPa/K, r is the psychrometer constant, kPa/K.

3 Study area and data sources

3.1 Overview of the study area

Zhejiang Province has a terrestrial area of 101,800 km^2, of which 70.4% are hills and mountains (Fig. 2). The ground elevation ranges from 0 to 1914m. It has mild temperatures, and vegetations grow well. In land-use types, forest land accounts for 62.8%, followed by paddy fields and towns. The average annual

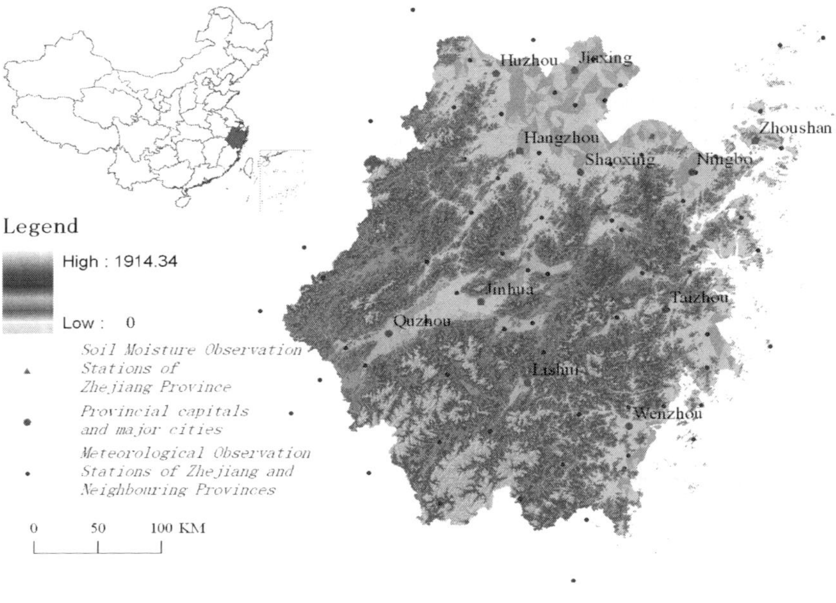

Figure 2: Topographic map of Zhejiang and sampling sites locations.

precipitation ranges from 980 mm to 2000mm. Affected by the plum rains and typhoons, the precipitation mainly occurs between April and October. The high-intensity rainfall and great elevation difference result in a frequent occurrence of landslide and debris flow.

3.2 Data sources

The MODIS instrument has moderate spatial resolution (250m, 500m and 1000m), high temporal resolution (four times per day), and high spectral resolution (36 spectral bands range from 0.62μm to 14.385μm).

The 6 days of MODIS data used in this paper were downloaded from the website of NASA. They were obtained on clear days from July to September, 2008. The data include surface reflectance product (MOD09), surface temperature product (MOD11) and normalized difference vegetation index products (MOD13).

Air temperature, wind speed and vapour pressure data of the satellite transit time were obtained from 72 meteorological stations in Zhejiang Province and 11 stations in the neighbouring provinces (Fig. 2). Soil moisture data were also obtained from 10 meteorological stations of Zhejiang Province for verification. In addition, land use classification maps and digital elevation model (DEM) data were used in this paper.

4 Data processing and TVDI and CWSI retrieving

4.1 Data processing

4.1.1 TVDI data processing

4.1.1.1 Surface temperature adjustment TVDI is based on the negative relationship between T_s and *NDVI*. An even and stable atmosphere is essential for the computation of this index. In this paper, T_s values need to be adjusted because atmosphere is non uniform in the study region. In the troposphere, air temperature drops 0.65 °C when the height increases by 100m. As a result, T_s values were adjusted as follows:

$$T_s' = T_s + 0.0065Z \tag{9}$$

where Z is ground elevation, m.

4.1.1.2 Determination of the dry edge and wet edge T_s' was plotted versus *NDVI* in Fig.3. We can see that T_s' increases with *NDVI* when *NDVI* is less than 0.3, which is contrary to the theory of feature space. It induces some difficulty to determining the dry edge. According to the land-use map, we found that pixels with *NDVI* values less than 0.3 are mostly mixed pixels. Few pure pixels can be found at the 1km pixel scale. It results in a lack of bare ground pixels. Therefore, we only used pixels with *NDVI* values greater than 0.3 to determine the dry edge. The maximum T_s' value was selected for each *NDVI* interval of 0.01, then they are used to fit the dry edge (see Table 1).

Table 1: Wet and dry edges for *Ts-NDVI* Feature Space.

Time	Dry edge	Wet edge
7/6/2008	y = -7.8671x + 48.259	y = 28.04
7/17/2008	y = -8.4107x + 45.676	y = 27.95
7/27/2008	y = -7.1457x + 41.768	y = 23.79
8/12/2008	y = -6.499x + 40.548	y = 25.77
9/10/2008	y = -3.579x + 37.467	y = 24.01
9/22/2008	y = 0.4521x + 35.907	y = 25.62

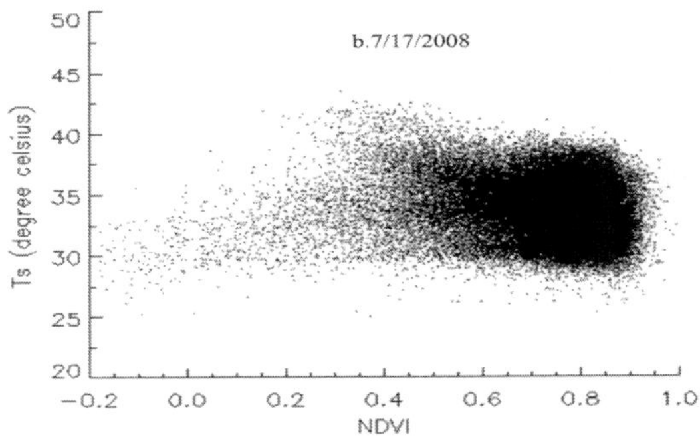

Figure 3: Scatter point TVDI image of Zhejiang (case study of July 17th 2008).

Soil moisture is high in the study area in summer because of plenty of precipitation, so we can get wet pixels for different *NDVI* values. For the selected 6 days, standard deviation of the minimum temperatures in each *NDVI* interval is 1.34 ℃, 1.28 ℃, 1.29 ℃, 1.50 ℃, 1.20 ℃ and 1.08 ℃ respectively. It proves the low discrete degree of the data. As a result, the wet edge is expressed by the average value (see Table1).

4.1.2 CWSI data processing

4.1.2.1 Input data of model Input data of the model include 3 categories: (1) Remote sensing data: surface albedo, *NDVI*, and surface temperature; (2) Meteorological data: air temperature, vapour pressure and wind speed; (3) Geographical data: altitude, slope, flow direction, latitude, and land use. These data are directly or indirectly used in the model.

4.1.2.2 Spatial interpolation of meteorological data We got the raster data of air temperature, wind speed and vapour pressure through the Kriging interpolation with single-station data. The resolution is 1km×1km. In order to eliminate the impact of elevation and make two-dimensional interpolation to one-dimensional, air temperatures were revised to the sea level temperatures according to Equation (9) before interpolation. They were revised to the actual altitude after interpolation.

4.2 Results of TVDI and CWSI retrieving

The results are retrieved by TVDI and CWSI (Fig.4-5). The majority values of results are between 0 and 1, and the region's values less than 0 are under the cloud coverage.

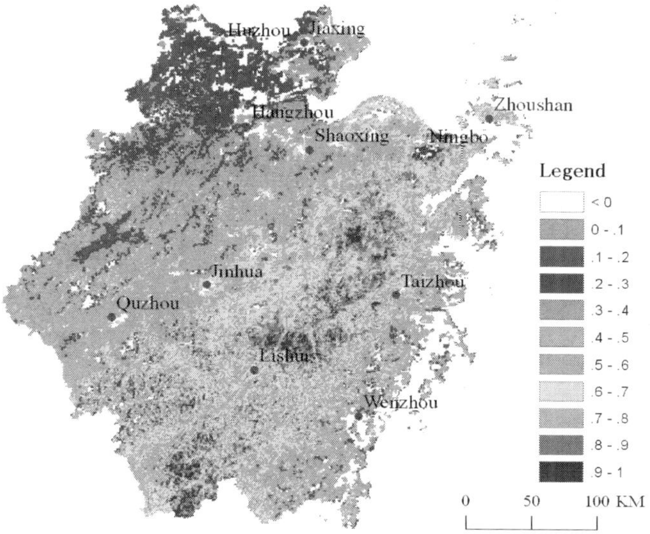

Figure 4: Retrieving results of TVDI (case study of July 17th 2008).

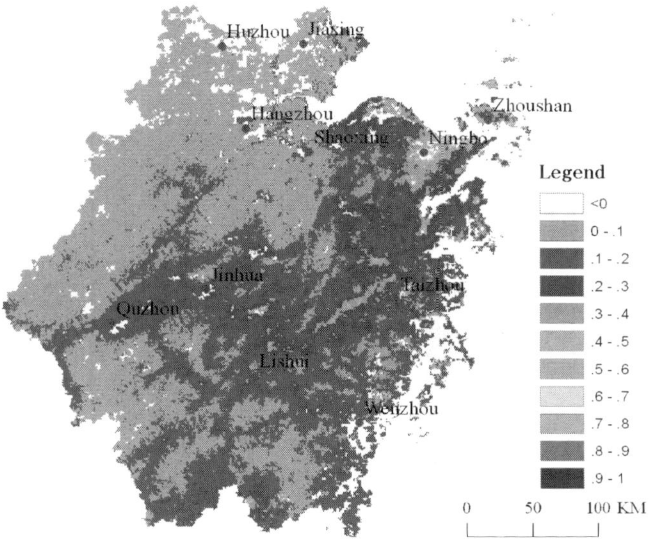

Figure 5: Retrieving results of CWSI (case study of July 17th 2008).

4.3 Examination and comparison of retrieving results

4.3.1 Examinations of retrieving results with observation data

We got soil moisture from 11 stations in the study area. 5 Urban stations were excluded during the data processing because of the complexity in land use as well as the mixed pixels of MODIS images. TVDI and CWSI were compared with relative soil moisture (RSM) observations at different depths, separately.

The correlation coefficients (Table 2) are negative. According to the average correlation coefficient, CWSI has a closer relationship with soil moisture than TVDI.

4.3.2 Comparison of retrieving results between two methods

In the study area, drought degree is defined as follows. RSM in 0%~20% represents extremely severe drought, 20%~40% represents heavy drought, 40%~60% represents slight and medium drought, 60%~90% represents suitable, and >90% represents extreme wetness. All of the RSM values used in this study are over 40%. Some of them even arrive at 100%. Generally, extremely severe drought and heavy drought didn't occur in the selected days.

Qi [16] defined five drought levels according to TVDI and four drought levels of CWSI, as listed in Table 3. When TVDI values in Zhejiang Province are classified with this method, results are quite different with the real conditions (Table 4). CWSI is more suitable for the actuality. Simulated values of CWSI are between 0-0.7. Table 4 lists mean values of the RSM observations corresponding to different CWSI intervals. It can be seen that CWSI can reflect the overall trend of soil moisture distribution.

Table 2: Correlation coefficients between RSM and TVDI and CWSI in different soil layers.

Soil depth	TVDI	CWSI
0-10cm	-0.74	-0.89
10-20cm	-0.62	-0.77
20-30cm	-0.80	-0.64
Average	-0.72	-0.77

Table 3: Drought classifications according to TVDI [16] and CWSI [17].

TVDI	CWSI*	Description
0～0.2	<0.4	Moisture
0.2～0.4	0.4～0.6	Normal
0.4～0.6	0.6～0.8	slight and medium drought
0.6～0.8	/	Drought
0.8～1	0.8～1	Severe Drought

Note: "/" No data in this range. "*" Classification refers to the Water Deficit Index (WDI) partition method based on the CWSI theory.

Table 4: Mean RSM values for different TVDI and CWSI intervals.

Intervals	Mean RSM values for TVDI			Mean RSM values for CWSI		
	0-10cm	10-20cm	20-30cm	0-10cm	10-20cm	20-30cm
0	/	/	/	90.7	92.0	95.0
0-0.1	/	/	/	73.5	84.2	81.4
0.1-0.2	/	/	/	/	/	/
0.2-0.3	/	/	/	76.8	78.6	75.6
0.3-0.4	90.7	92.0	96.0	49.5*	60.8*	67.5*
0.4-0.5	59.0*	63.0*	78.0*	50.0	77.0	79.0
0.5-0.6	72.8	81.6	82.2	/	/	/
0.6-0.7	74.2	86.8*	80.6	/	/	/
0.7-0.8	56.2	70.4	72.9	/	/	/
0.8-0.9	41.0	54.0	55.0	/	/	/
0.9-1	/	/	/	/	/	/

Note: "/" No observations in this range. "*" The value is small due to lack of samples.

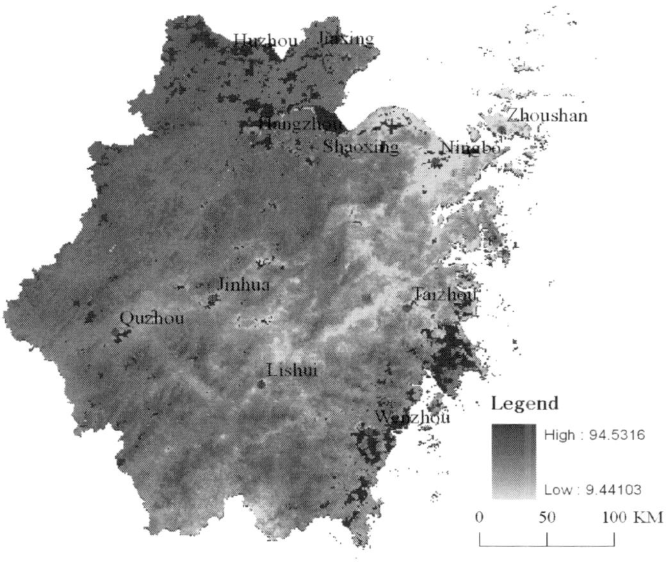

Figure 6: Estimated 0-10cm RSM of Zhejiang on July 17th 2008 using correlation equation $y_{0\text{-}10cm} = -93.586x + 85.173$.

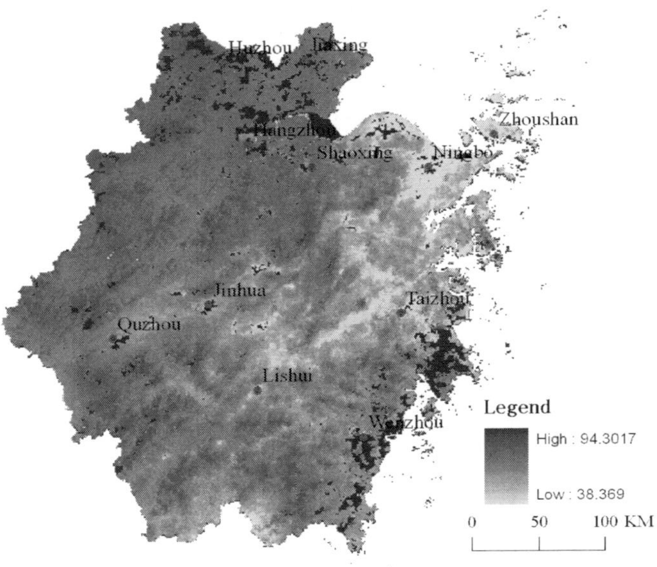

Figure 7: Estimated 20-30cm RSM of Zhejiang on July 17th 2008 using correlation equation $y_{10\text{-}20cm} = -61.517x + 88.15$.

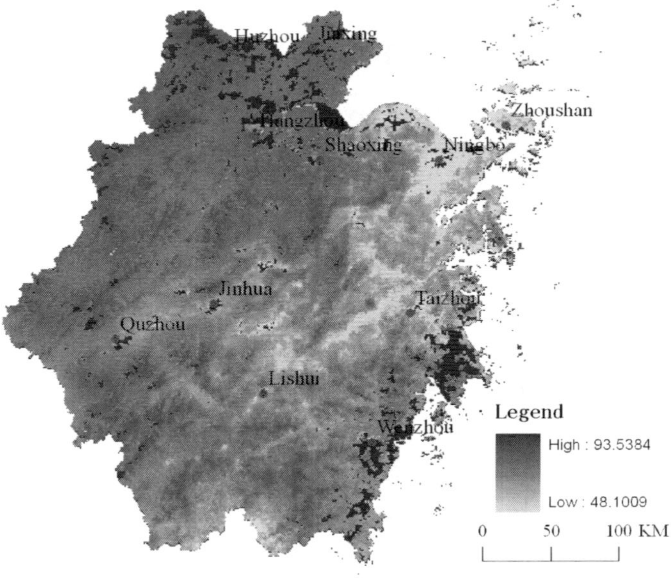

Figure 8: Estimated 20-30cm RSM of Zhejiang on July 17th 2008 using correlation equation $y_{20\text{-}30cm} = -49.974x + 88.541$.

4.4 Soil moisture retrieval using CWSI

CWSI has a closer relationship with soil moisture than TVDI, so it was selected to retrieve RSM in Zhejiang Province. Both fitting equations between CWSI and RSM observations and calculated RSM sketch maps are shown in Figure 6-8 for each soil layer. As illustrated, RSM shows higher in south, west and north, while lower in central and east on July 17, 2008. Three levels of soil moisture distribution have the same trend. The RSM will further provide the basis for debris flow forecast.

5 Conclusions

In this paper, MODIS data were used to retrieve soil moisture in Zhejiang Province, China. In humid regions with high vegetation cover, the experiential dry edge of TVDI may be not the actual dry edge. It can lead to higher estimation in TVDI and severer extent in draught judgement. So TVDI cannot be directly used to express draught degree in this area. However, TVDI has indeed closely relationship with soil moisture, and TVDI does not require meteorological data. If we can find actual dry edge in humid region, it may have more broad application prospects.

CWSI simulated with SEBS can be used in humid regions with high vegetation cover and has more potential to retrieve soil moisture, because SEBS has more advantages in calculating evapotranspiration. It is more objective to determine CWSI because wet and dry edges are not required. CWSI can better reflect soil moisture conditions, especially in the surface layer.

Soil moisture retrieval with remote sensing images resolves the uncertainty with antecedent rainfall and provides an important basis for debris flow forecast. It has far-reaching significance in debris flow prediction.

Acknowledgements

This research was supported by the Research Fund for Commonweal Trades (Meteorology) (GYHY200706037) and the Key Project in the National Science & Technology Pillar Program (Grant number: 2008BAK50B04).

References

[1] Wei, F; Gao, K; Cui, P, et al. Method of debris flow prediction based on a numerical weather forecast and its application. In: Lorenzini G; Brebbia CA; Emmanouloudis D (eds) Monitoring, Simulation, Prevention and Remediation of Dense and Debris Flows: WIT Press, Ashurst Lodge, Southampton, pp.37-46, 2006.

[2] Wei, F; Hu, K; Zhang, J, et al. Determination of effective antecedent rainfall for debris flow forecast based on soil moisture content observation in Jiangjia Gully, China. In: DeWrachien D; Brebbia CA; Lenzi MA (eds) Monitoring, Simulation, Prevention and Remediation of Dense Debris Flows II: WIT Press, Ashurst Lodge, Southampton, pp.13-22, 2008

[3] Sandholt, I., Rasmussen, K. et al. A simple interpretation of the surface temperature/vegetation index space for assessment of surface moisture status. Remote Sensing of Environment 79: pp. 213-224, 2002.
[4] Idso, S., Jackson, R. et al. Normalizing the stress-degree-day parameter for environmental variability. Agricultural Meteorology: 24, pp. 45−55. 1981.
[5] Jackson, R., Idso, S. et al. Canopy temperature as a crop water stress indicator. Water Resources Research: 17, pp.1133−1138. 1981
[6] Carlson, T. N., Gillies R. R., et al. A method to make use of thermal infrared temperature and NDVI measurements to infer surface soil water content and fractional vegetation cover. Remote Sensing Review: pp.161-173, 1994
[7] Nemani, R. R., Pierce, L., et al. Developing satellite-derived estimates of surface moisture status. Journal of applied Meteorology 32 (2): pp. 548-557, 1993.
[8] Price, J. C. Using spatial context in satellite data to infer regional scale evapotranspiration. [J]. IEEE Transactions on Geoscience and Remote Sensing (28): pp. 940-948, 1990.
[9] Claps, P. and Laguardia G. Assessing spatial variability of soil water content through thermal inertia and NDVI. Remote Sensing for Agriculture, Ecosystems, and Hydrology V 5232: pp. 378-387, 2004.
[10] Mallick, K., Bhattacharya, B. K., et al. Evapotranspiration using MODIS data and limited ground observations over selected agroecosystems in India. International Journal of Remote Sensing 28(10): pp. 2091-2110, 2007
[11] Patel, N. R., Anapashsha, R., et al. Assessing potential of MODIS derived temperature/vegetation condition index (TVDI) to infer soil moisture status. International Journal of Remote Sensing 30(1): pp. 23-39, 2009.
[12] Su Z. The Surface Energy Balance System (SEBS) for estimation of turbulent heat fluxes. Hydrology and Earth System Sciences: 6(1), pp.85-99, 2002.
[13] He, Y., Su Z, Jia L, et al. Estimation of Surface Energy Flux Using Surface Energy Balance System(SEBS)in the Yellow Huaihe-Haihe River Regions, China. Plateau Meteorology: 25(6): pp. 1092-1100, 2006. (in Chinese)
[14] Jin, X., Tang, Y., Research on Regional Evaportranspiration of Three Basins in Shanxi Based on Remote Sensing Method. Science & Technology Review: 25(4): pp. 31-34, 2007. (in Chinese)
[15] Moran, M. S., Clarke, T. R., et al. Estimating Crop Water Deficit Using the Relation between Surface-Air Temperature and Spectral Vegetation Index. Remote Sensing of Environment: 49, pp. 246-263, 1994.
[16] Qi, S., Drought Monitoring Models with Remote Sensing and Spatio-Temporal Characteristics of Drought in China. PH.D Dissertation. Institute of Remote Sensing Applications, Graduated University, Chinese Academy of Sciences.2004.(in Chinese)
[17] Hu, M., The research on the drought and its influence on the production of grain in Jiangxi province by Remote Sensing. Master's Thesis. Jiangxi normal University.2008.(in Chinese)

Debris flow induced by glacial lake break in southeast Tibet

Z. L. Cheng[1,2], J. J. Liu[1,2,3] & J. K. Liu[1,2,3]
[1]*Key Laboratory of Mountain Hazards and Surface Process, Chinese Academy of Sciences, China*
[2]*Institute of Mountain Hazards and Environment, CAS, China*
[3]*Graduate University of Chinese Academy of Sciences, China*

Abstract

Debris flow caused by flooding of a glacier-lake break can result in major disasters in southeast Tibet. Based on field investigation and data analysis, this paper discusses the influences of climate factors on lake break. It is found that the break is inclined to occur after abnormal changes in climate, particularly at the turn from cold-wet weather to hot-dry weather. Debris flows due to glaciers will be more active in future decades.

Keywords: debris flow induced by glacier-lake break, climate change, disaster, southeast Tibet.

1 Introduction

Debris flow induced by glacial lake break has brought about tremendous disasters to lives and properties in southeast Tibet. For example, icefalls occurred widely on June 29, 1983, August 23, 1984, and July 20, 1985, blocking the glacier and forming temporary lakes. All three lakes finally broke and the flood created great debris flows, with peak discharge up to 2950, 5245, and 8195 (m^3/s), respectively. They blocked the south Sichuan-Tibet highway for 270 days, destroyed 54 bridges, 8hm^2 farmland, 22 houses, and 79 trucks, amounting to a loss of 2 million dollars (Fig.1). another event is the break of the Guangxiecuo lake, where the resulting debris flow entered the upper Palongzangbu River at a discharge of 1021.4m^3/s and partially blocked the river, raising the water level up to more than 10m. The flow rushed down, destroying 42km highway, and suspending traffic for 200 days. It also washed out 18

WIT Transactions on Engineering Sciences, Vol 67, © 2010 WIT Press
www.witpress.com, ISSN 1743-3533 (on-line)
doi:10.2495/DEB100091

bridges, destroyed 51 houses, and killed 5 persons. The total loss amounted to more than 10 million dollars [1]. Since the event this section has become more and more fragile and suffered from frequent traffic accidents [2]. Glacial lakes are mainly distributed in the rivers of Yaluzangbu, Palongzangbu, Niyang, Boqu, and Pengqu (Fig.2), and control measures have been taken since the 1980s [3–8].

Figure 1: A debris flow fan in Peilong Gully.

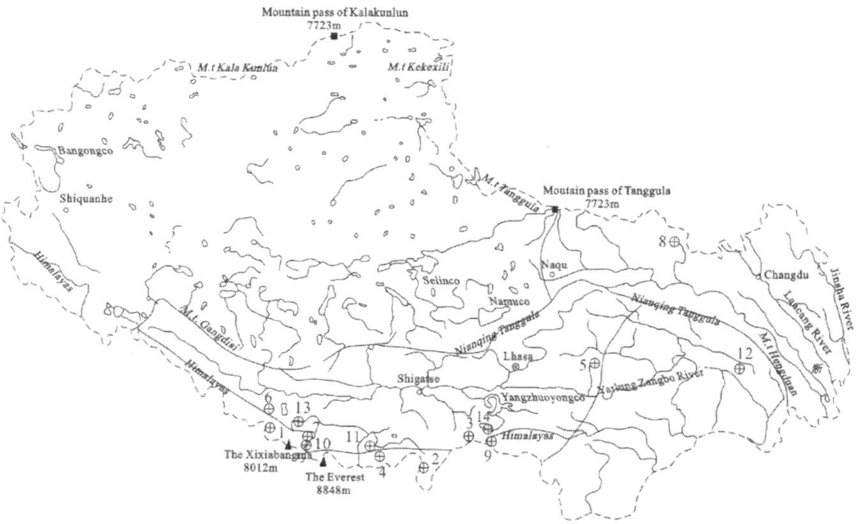

Figure 2: The distribution of debris flow induced by GLOFs in Southeast Tibet (excluding the lakes resulting from icefall), [9]: 1—Taco, 2—Qiongbihemaco, 3—Sangwangco, 4—Jilaico, 5—Damenlakeco, 6—Longdaco, 7—Ayaco, 8—Bangeco, 9—Zharico, 10—Cirenmaco, 11—Jingco, 12—Guangxieco, 13—Jialongco, 14—Degaco.

The climate is crucial to the outburst of glacial lakes. Tufnell [10] analyzed more than 50 events in the Alps, finding that over 95% happened between July and September and were concentrated in June and August. Out of the 18 identified events in Tibet, six occurred in August and in the extreme alpine regions of the Himalayas [11]. Climate change may be the cause of outbursts throughout the world. The events in Tibet mainly happened in 1964, when the weather fluctuated greatly between cold and hot; and also in the transition from the cold-wet years of the 1960s to the hot-dry years of the 1970s and 1980s. In particular, abnormal weather may be the direct cause of lake break. It is reported that the event at Guangxiecuo was due to the persistently high temperature and extreme rainstorm.

This paper discusses the climate influence on glacial lake break, and analyses the trend of glacial lakes in southeast Tibet under global warming conditions.

2 Climate factors

Many factors are influential on the breaking of glacial lakes, including the size and thickness of glaciers, the area and impounded water of the lake, structure and stability of the embankment, and temperature and precipitation. Among them the last two are decisive. In the monsoon climate the summer is warm and rainy in southeast Tibet: this increases the melted water, which enters the lake and raises the water level or even brings about overflow. Heavy rainfall is also important. On the other hand, the melted water converges toward the glacial front and infiltrates into the cracks. It fabricates the substrate and fills the pores. A critical equilibrium occurs when the captured water reaches 0.9 times the original depth. Once the equilibrium breaks, i.e., the shear stress of the front goes beyond the friction between the glacier and the substrate, the front will jump and the glacier splits into pieces of icefalls pouring into the lake. Then the water rises and the wave surges and wallops the moraine dike and finally causes the lake collapse.

The 25 break events display a close relationship to the climate condition. All the events happened between May and September, particularly June and August (Table 1). Precipitation is abundant in April and July in southeast Tibet, which provides plenty of water for the break in the following months.

3 Annual variation of climate and glacial lake break

Climate conditions change over months and years, as does the breaking of glacier lakes.

Table 1 and fig.4 indicate that the rate of break is two or less for the ten years before 1960, and for 1970–1980 and 1990–2000; and it is seven or less for 1960–1970 and 1980–1990. The present decade is another active period, seeing four events from 2000 to 2008.

Table 1: The date of debris flow due to glacier lake break in southeast Tibet [9, 11].

Glacier lake	Watershed	Country	Burst date	Disaster forms	The causes for outburst
Lulang lake	Polong zangbu	Bomi	1931.6.8	Dilute debris flow, floods	Icefall, Glacial movement
Daruoco	Boqu	Nielamu	1935.8.28	Floods, debris flow	Glacial movement
Qiongbihe maco	Kangbuqu	Yadong	1940.7.10	Floods, debris flow	Icefall
Sangwangco	Nianchu	Kangma	1954.7.16	Dilute debris flow, floods	Glacial movement
Jilaipuco	Pengqu	Dingjie	1964.9.21	Dilute debris flow, floods	Glacial movement
Longdaco	Jilong zangbu	Jilong	1964.8	Floods, debris flow	Icefall, Glacial movement
Damen lakeco	Niyang	Gongbu jiangda	1964.9.28 1968.8.15	Floods, Dilute debris flow, Viscous debris flow	Icefall, Glacial movement
Ayico	Pengqu	Dingri	1969.8.17 1970.8.17	Floods, Dilute debris flow, Viscous debris flow	Glacial movement
Bangeco	Nujiang	Suo	1968.8 1972.7	Dilute debris flow, floods	Icefall, Glacial movement
Zharico	Luozha xiongqu	Luozha	1981.6.24	Floods, debris flow	Icefall, Glacial movement
Zhang zangco	Boqu	Nielamu	1964	Floods, debris flow	Glacial movement
Ciren maco	Boqu	Nielamu	1981.7.11	Floods, debris flow	Icefall, Piping
Jingco	Pengqu	Dingjie	1982.8.27	Floods, debris flow	Icefall
Upstream Peilong	Palong zangbu	Linzhi	1983.7.29 1984.8.23 1985.6.20	Floods, Viscous debris flow	Icefall

Table 1: Continued.

Guangxieco	Palong zangbu	Bomi	1988.7.14	Dilute debris flow, floods	Icefall, Glacial movement
Jianglong lake	Boqu	Nielamu	2002.5.23 2002.6.29	Viscous debris flow, Dilute debris flow, Floods	Icefall, Glacial movement
Degaco	Luozha xiongqu	Luozha	2002.9.18	Viscous debris flow, Dilute debris flow, Floods	Icefall, Glacial movement
Jianmu puqu	Palong zangbu	Linzhi	2008.4.9	Viscous debris flow, Dilute debris flow, Floods	Icefall

Table 2: The monthly mean temperature at the entrance to Peilong Gully (1984m) during 1982–1985 (°C).

month / year	1	2	3	4	5	6	7	8	9	10	11	12	annual
1982	5.0	6.1	9.2	13.1	16.4	18.8	21.1	21.8	18.4	13.5	8.2	5.1	13.1
1983	3.1	4.8	8.9	11.9	17.0	19.1	21.4	21.0	19.0	14.4	8.2	4.5	12.8
1984	4.7	8.4	11.1	13.2	16.5	19.0	20.6	20.9	17.9	15.5	8.0	6.5	13.6
1985	5.2	7.2	10.9	13.6	16.8	19.0	20.1	22.0	18.3	13.4	8.6	5.9	13.4
average	4.5	6.6	10.0	13.0	16.7	20.0	20.8	21.9	18.4	14.2	8.3	5.5	13.2

Table 3: The monthly precipitation at the entrance to Peilong Gully (1984m) during 1982–1985 (mm).

month / year	1	2	3	4	5	6	7	8	9	10	11	12	annual
1982	8.6	46.6	110.2	171.9	144.4	357.5	136.7	40.8	314.4	84.3	45.4	18.9	1497.7
1983	9.4	61.0	146.3	127.0	67.0	192.3	138.0	63.1	114.9	144.9	27.5	8.9	1100.3
1984	11.0	17.7	106.3	220.8	104.1	243.4	157.3	122.0	145.0	116.6	10.3	21.4	1257.9
1985	21.0	13.5	188.4	244.5	109.6	319.9	182.6	87.8	181.6	69.4	16.9	36.8	1472.0
average	12.5	34.7	137.8	191.1	106.3	278.3	153.7	78.4	189.0	103.8	25.0	21.5	1332.0

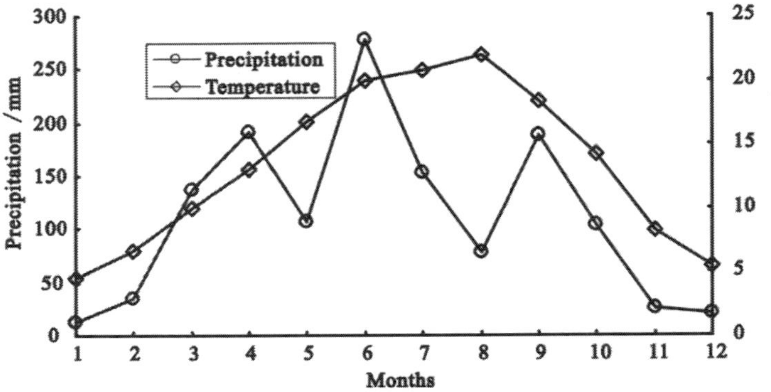

Figure 3: Monthly mean precipitation and temperature during 1982–1985 in the outlet to Peilong Gully, Linzhi County.

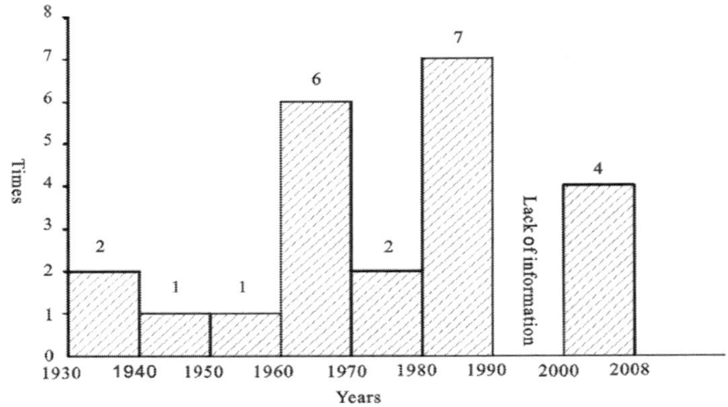

Figure 4: Annual variation of glacier lake outburst in Southeast Tibet.

3.1 Warming and lake growing

While cold-wet weather facilitates growth and advance the glacier, hot-wet and hot-dry weather result in melting and retreat, and therefore increase the number and size of glacier lakes [12].

The regional warming of Tibet in the last 50 years, with remarkable annual fluctuation of rainfall (Figs.8, 9), has increased both the number and size of glacial lakes. For examples, the increase is up to 11% and 47%, respectively, for the number and area of glacial lakes in the Boqu River, from 1987 to 2005 (11 lakes with 17.6km^2 in area). In Pengqu, the area has increased by 13% since the 1980s, amounting to 47.5km^2, and in Dingjie County, the lake has grown by 0.12km^2 and the water has risen by 1m, with the impounded water up to 5.5million m^3.

Another example is the Palongzangbu, where many large monsoon oceanic glaciers have developed. According to Landsat TM data of 1990, 2005, and

2006, there are 732 glacier lakes in this area, out of them 142 were created between 1970 and 2006, amounting to 10.4km^2. Figs.5–7 show three examples.

Using available data we have analyzed changes to 493 glacial lakes in the years 1970, 1990, and 2000. The results show that the lake area increased by 9.11% during 1990 and 2000; and 215 (including 75 newly born) lakes have grown since 1970. There are 99 lakes whose area has doubled or more. There are also 278 lakes that have shrunk or even disappeared (31). Some lakes have

Figure 5: Duoyico, taken by Xie Hong.

Figure 6: Kangzhaico, taken by Xie Hong.

Figure 7: Guangxieco, taken by Xie Hong.

increased remarkably in recent years, including Dankalongba in the south of Bomi, 10km from the Palongzangbu, which were measured at 206178m^2 and 214222m^2, respectively, in 1987 and 1999, while they were absent in 1970s. Guangxieco, with only 0.08km^2 left after the break in 1988, is 0.20km^2 at present. There are other lakes with an area increased by 0.2km^2.

3.2 Abnormal climate and lake break

In the early 1960s glaciers in Tibet grew widely because of the cold-wet weather. However, they are prone to break as the weather becomes hot and dry. Climate abnormality is particularly related to the lake failure. Abrupt changes in climate occurred in the early 1960s and early 1980s (Table 4). A comparison between Table 1 and Table 4 indicates that lake failure occurred mainly in the abnormal years, such as the events in 1964, 1968, 1969, and 1970. The seven occurrences in the 1980s also followed the climate change.

Table 4: The years of abnormal climate between 1952 and 1995 [13].

	Years of close to abnormal climate	Years of abnormal climate
Summer	1961(+) 1965(-) 1972(+) 1981(+) 1983(+) 1986(+) 1994(+) 1995(+)	1968(+) 1976(-) 1977(-)
Autumn	1974(+)	1964(+) 1965(-) 1967(-)

Note: (+) indicating warming; (-) indicating cooling.

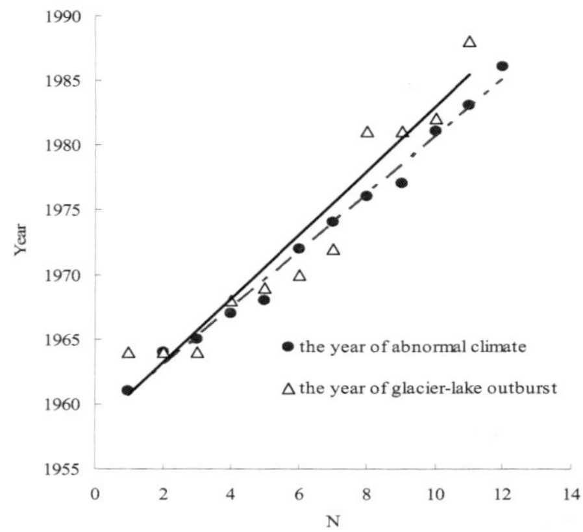

Figure 8: Correlation between the occurrences of glacier-lake break and abnormal weather.

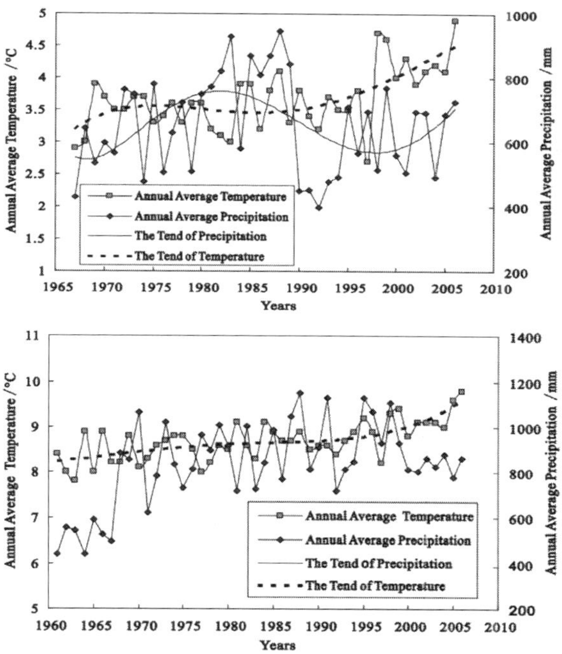

Figure 9: Annual variation of glacier lake outburst in southeast Tibet.

Comparing the abnormal years and the break events, a closed relationship appears:

$$\text{Temperature abnormal year: } T=2.2168N+1958.4 \quad (R^2=0.9902)$$

$$\text{Break events: } T=2.4727N+1958.2 \quad (R^2=0.9239)$$

where N is the order of break event and T is the corresponding year. The break line has a slope bigger than that of the temperature line, suggesting that there is a lag for the effect of temperature change. Break occurs after the climate change.

4 Tendency of the glacial lakes

It is expected that, compared with 1990, Tibet will have a rise of temperature of 0.8–1.2°C and rainfall of 7–17% by 2030; and the rise will be much higher by 2050. Fig.9 indicates that the temperature has been continuing to rise since 2000. In Bomi County, the temperature reached 9.8°C in 2006 and the increase rate is 12.7%; the rate is even as high as 33.5% in Nielamu county, where the temperature in 2006 was 4.9°C, while the average before that year was 3.67°C. Precipitation in these regions is rising with fluctuation. It is predicted that the runoff of the melted water will increase by 50%, while the glacier area will shrink by half.

The increase of temperature and rainfall, companied with the extreme weather conditions such as rainstorms, will certainly favour breaking of glacier lakes, which will in turn result in debris flow because of the accompanied increase of loose material supplies in the valleys. It is concluded that debris flow due to the breaking of glacier lakes will be more active in the coming decades in southeast Tibet.

On the other hand, the glacier area will decrease with the rise of temperature. When the glacier area shrinks too much the melted water will not be enough to result in a lake break. There is a critical state of glaciers concerning the lake break. Finding the state should be an interesting study for the future.

5 Proposal for countermeasures

Because the debris flows induced by breaking of glacier lakes are always on a large scale, the normal countermeasures for control, such as the sediment dam and drainage groove, are highly expensive but not sufficiently powerful. So, the focus must be put on controlling the dynamical conditions that facilitate the formation of debris flow and on preventing the break of glacier lakes. Glacier lakes, particularly the end-moraine lakes, are a special kind of dammed lake that is similar to the dammed lakes of landslides and therefore can be treated with the same countermeasures [14, 15].

There have been various methods treating moraine lakes in the world. The simplest one is to excavate a groove to drain the water and reduce the water level. This has been widely used in treating landslide lakes, such as the dammed lakes in the Wenchuan Earthquake in 2008; and it has been also used in treating moraine lakes, such as in Peru [16]. In addition, to reduce the power of the overflow, dams can be built in the gentle section downstream from the lake.

Acknowledgements

This research is supported by the National Natural Science Foundation of China (Grant No. 40771024), Projects in the National Science & Technology Pillar Program (2008BAKSOBO4), and the Non-profit Project of the Ministry of Water Conservancy (200801032).

References

[1] Zhu Pingyi, He Ziwen, Wang Yangcun et al. The study of typical mountain disasters in Sichuan-Tibet Highway. Chengdu: Chengdu science and technology university Press, pp.35–156, 1999

[2] Lv Ruren, Li Deji. Debris flow induced by glacier lake outburst in Tangbulang gully, Gongbujiangda, Tibet. Journal of Glaciology and Geocryology, 8(1), pp. 61–64, 1986.

[3] Xu Daoming. Characteristics of debris flow caused by outburst of glacier lake on Boqu river in Xizang. Journal of Glaciology and Geocryology, 9(1),pp.23–24, 1987

[4] Wang Tiefeng, Liu Zhirong, Xia Chuanqing, *et al.* Study of glacier lakes in Nianchu river basin, Tibet. Journal of Glaciology and Geocryology, 25(suppl.2), pp.344–348, 2003.
[5] Chen Chujun, Liu Ming, Zhang Zhi. Outburst conditions of moraine dammed lakes and their flood estimation in the headwaters of Nianchu river, Tiber. Journal of Glaciology and Geocryology, 18(4), pp.347–352, 1996.
[6] Chen Xiaoqing, Chen Ningsheng, Cui Peng. Calculation of discharge of debris flow induced by glacier lake outburst. Journal of Glaciology and Geocryology, 26(3), pp.357–362. , 2004
[7] Chen Xiaoqing, Cui peng, Li Yong, et al. Changes of glacier lakes and glaciers of post-1986 in Bioqu river basin, Nyalam, Tibet. Geomorphology, 24(5), pp.298–311, 2008
[8] Cui Peng, Ma Dongtao, Chen Ningsheng, et al. the initiation, motion and mitigation of debris flow caused by glacier lake outburst. Quatermary sciences, 26(3), pp.621–628, 2003
[9] Liu Jingjing, Cheng Zunlan, Li Yong et al. Characteristics of Glacier. Lake Breaks in Tibet .Journal of catastrophology. 23(1), pp. 54-60, 2008
[10] Tufnell, L.: 1984, Glacier Hazards, Topics in Applied Geography, Longman, New York.
[11] Chinese Academy of Sciences - Ministry of Water Resources Chengdu Institute of Mountain Hazards and Environment, Department of the Tibet Autonomous Region Institute of Science, mud-rock flow in Tibet and the environment, Chengdu Science and Technology University Press, pp.106–136 ,1999
[12] Wang Jiewu, Dong Guangrong .Assessment of evolution of western China Environmental (Volume I: Characteristics and evolution of the environment in western China. Beijing: Science Press, pp. 50–51, 2002
[13] Du jun. Change of temperature in Tibetan Plateau from 1961 to 2000. Acta Geographica Sinica, 56(6), pp. 682-690., 2001
[14] Ding Yihui. Assessment of environment in west China. Vol.2, Forecasting of environment changes in west China .Beijing: Science Press, pp.39, 2002.
[15] Che Tao, Jin Rui, Li Xin, et al. glacier lakes variation and potentially dangerous glacier lakes in Pumqu river basin in Tibet during of last two decades. Journal of Glaciology and Geocryology, 26(4), pp.397–402, 2004.
[16] Qing Dahe. Assessment of environmental changes in west China. Beijing: Science Press, pp.61–65, 2002

Experience with treatment of road structure landslides by innovative methods of deep drainage

O. Mrvík[1] & S. Bomont[2]
[1]*Department of Geotechnics, Czech Technical University in Prague, Czech Republic*
[2]*TP.GEO, France*

Abstract

During construction of infrastructure projects, emergency situations due to the presence of groundwater and slope instability occur very often. In addition, damage to existing roads or railways caused by groundwater or slope deformations do not represent any extraordinary situation. In such cases, the water should be taken from the ground in order to improve the properties of the soils and rocks. According to the consumption of energy, the methods of dewatering can be divided into gravity dewatering and dewatering with electricity. Traditionally used drainage techniques are proven methods. However, in certain geotechnical conditions, two innovative drainage systems, Siphon Drains and Electropneumatic Drains, can prove to have many advantages. In this paper, several applications of the innovative deep drainage systems are described. The paper introduces selected sites where groundwater lowering played a key role in the successful solution of slope stabilization and road remediation. The sites are located in France.

Keywords: road structures, landslide, groundwater, deep drainage, Siphon Drains, Electropneumatic Drains.

1 Introduction

Road structures, as roads, highways or railways, are very often endangered or even seriously damaged by different kinds of deformations of natural slopes or artificial excavations. The main trigger for such deformations lies in high groundwater levels within the affected area or just the subsoil of the road

WIT Transactions on Engineering Sciences, Vol 67, © 2010 WIT Press
www.witpress.com, ISSN 1743-3533 (on-line)
doi:10.2495/DEB100101

embankment. Therefore, groundwater lowering is primarily the most important measure in order to avoid the risk of slope movements, damage to the road structures and, secondarily, to protect the affected areas against repeated deformation and destruction.

Geotechnical designers responsible for the treatment of such problems should always recognize the main reasons for the situation, evaluate the geotechnical conditions of the site and look for the most effective solution. The decision-making process of a proper drainage system to drain and stabilize the affected site might be as follows.

Traditional methods of dewatering, such as drainage trenches, are usually limited by their maximal economical depth of 3–5 m. Deeper excavations would cause huge and non-economical earthworks and extra expenses for gravel backfilling. In the case of subhorizontal wells, considerable length of drilling, difficulties in reaching all aquifers and problems of site access can be considered as significant disadvantages. Alternatively, the innovative method of Siphon Drains®, which allows dewatering up to 8–12 m beneath the surface without the need for electrical energy, can be adopted.

Deep drainage and groundwater lowering deeper than 10 meters today still represents a great deal of effort and implies a difficult technique. However, sometimes it is inevitable to reach deeper aquifers that might cause many serious problems in construction and to deal with groundwater lowering of tens of meters. Traditional techniques are very often badly fitted to achieve such requirements. Submersible pumps require a rather high minimum flow and frequent expensive maintenance. Well points efficiency is limited to a depth of 7 meters lower than the level of the vacuum pump. The innovative method of Electropneumatic Drains® has been developed to lower the water table up to 40–60 meters below ground level using pneumatic energy (compressed air).

The groundwater problem at the first three of four introduced sites was treated by Siphon Drains as a permanent energy-free solution. The last case study introduces an application of Electropneumatic Drainage as a permanent solution to an emergency problem that occurred soon after construction of a project of highest importance.

2 Innovative methods of deep drainage

2.1 Siphon drains

Small diameter (10–30 mm) suction siphon tubes are placed into vertical wells. The wells are spaced between 3–6 m and are sufficiently deep to provide required drawdown. The wells are dewatered using the siphon principle based on gravity drawdown up to depths of 8–12 m beneath the surface (Fig. 1).

The tubes are inserted into a permanent water filled reservoir at the base of each well with an outlet downstream at an outlet manhole, situated down slope. If the water level rises in the well, the flow in the tube is renewed and abstracts water out of the well. The flow continues until the water level in the well falls back to the reference level, providing that the flow rate in the siphon is sufficient to keep the siphon primed. As the water rises towards the top of the tube, the

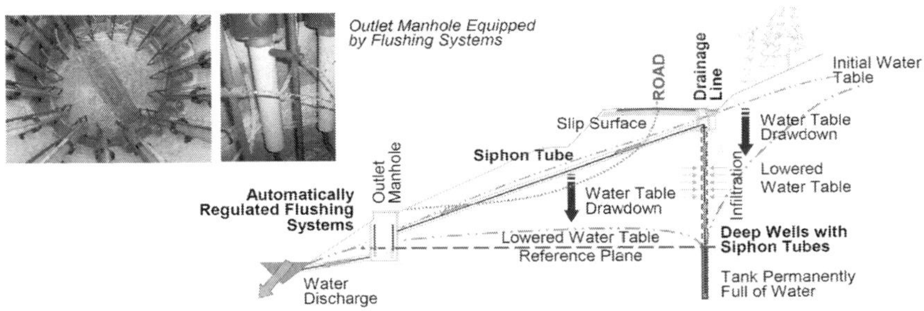

Figure 1: Basic principles of Siphon Drainage.

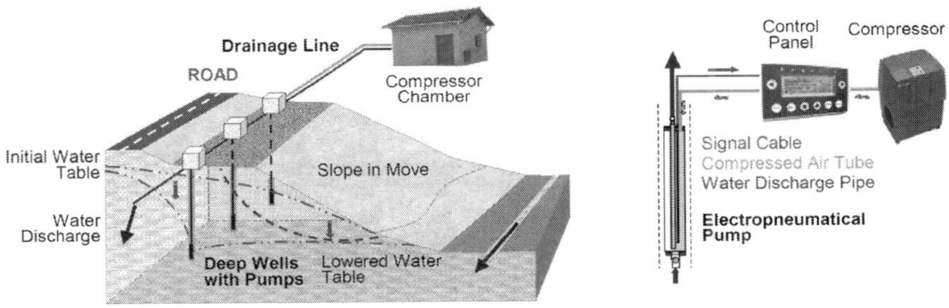

Figure 2: Arrangement of Electropneumatic Drainage.

pressure falls and may reach vacuum, causing the creation of bubbles. Without any additional measure, the bubbles might cause the flow to break. This is avoided by using the flushing system, which flushes bubbles out by turbulent flow and controls and regulates the flow so that the siphon always stays primed. The flushing system is an arrangement of PVC pipes and it is placed downstream in the outlet manhole at the end of each siphon tube.

The method is suitable for the geological environment of permeabilities less than 10^{-5} m/s and groundwater inflows of 0.0–2.0–15.0 l/min per well (i.e. 0.0–0.03–0.3 l/s per well). The main advantages are: the system is automatically continually in service, it is easily controlled, of high efficiency and can facilitate drawdowns of up to 12 m, without the need for any additional energy.

2.2 Electropneumatic Drains

The drainage is designed by a network of vertical wells, manholes, ducting for cabling and pipes for water discharge and compressed air (Fig. 2). The wells are spaced between 3–6 m and are sufficiently deep to provide the required drawdown. The wells are equipped with a casing of 110 mm diameter and a gravel filter. Electropneumatic pumps are installed into the wells at a defined depth, connected to a compressor and the control panel and equipped with an intelligent sensor. When the groundwater rises in the wells, it fills the pumps and when it reaches the high level sensor, an electrical signal is transmitted to the control panel, including the relays and solenoids. The signal triggers the injection of compressed air into the pumping chamber to push water out onto the surface through an outlet tube. The pump filling and emptying is repeated until the required drawdown is reached.

The system manages groundwater lowering of up to 60 m under permeabilities of 1.10^{-5}–1.10^{-7} m/s with groundwater inflows of 0.0–35.0 l/min per well (i.e. 0.0–0.6 l/s per well). The main advantages are: the system runs only under high groundwater level, the pumps contain no moving mechanical parts and both operation and maintenance expenses are saved.

3 RD74 – Goncourt

At this site, the affected section of the road is situated on a slope above a shallow valley with a river at the bottom (Fig. 3). The valley is filled mainly by soft sediments, such as clays. The bedrock is created by alternating marls and limestones. At the top of the slope, the limestones are exposed either by natural outcrops and cuttings made during construction of the road. The limestones are affected by karstic effects and by mechanical weathering. A complicated system of the groundwater of many aquifers confined mainly in cracks and voids is developed within the limestone formation. Infiltration of surface water was allowed. The water flow follows the slope so that the soft sediments in the lower parts of the valley were saturated. The water that occurred in the clays was very shallow, 0–5 m below surface. As the slope is being undercut by the river erosion, deformations in the saturated clays were initiated easily. These deformations were accompanied by gravity moves of limestone blocks in the upper part of the slope. Downstream movements of the limestone blocks caused deformations of the road. The affected area was 300 m in width and 150 m in length. The deepest deformations were expected to be 10 m.

To eliminate the deformations and to avoid future road destructions, the water coming from the top of the slope had to be drained so that the shear properties of the clays in which sliding occurred would be improved. Predominantly, long-term permanent gravity drainage without the need for energy was required. Pumping of the water from deep wells situated at the top of the slope would have been possible, but very time consuming. Continual water flow of a huge amount was expected. Horizontal wells carried out from the bottom of the slope would have to be very long. Reaching all of the required aquifers would not be guaranteed by horizontal boreholes. A drainage trench situated below the top of the hill was taken into account. Such a trench would have to be very deep to attain the deepest aquifer. The earthworks and the gravel backfill of such an excavation would be extremely expensive. Moreover, it might be dangerous to cut the slope by a 5–10 m deep trench that could break the weak stability or suddenly bring a lot of groundwater from cracks to the slope. The system of Siphon Drains was selected as the most suitable and effective.

A single drainage line of 71 drains was situated in the slope as an artificial water barrier (Fig. 4). The purpose of the drainage was to take the incoming water out of the slope and to lower the water table permanently. A shallow trench of 1.2 m minimum depth in which to place all necessary ducting, as well

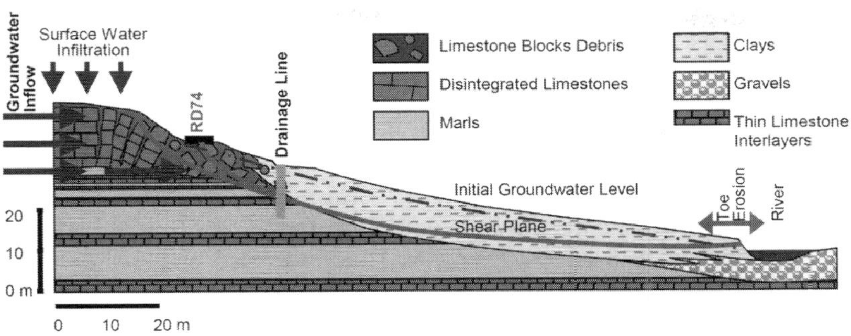

Figure 3: Cross-section of the affected slope at Goncourt.

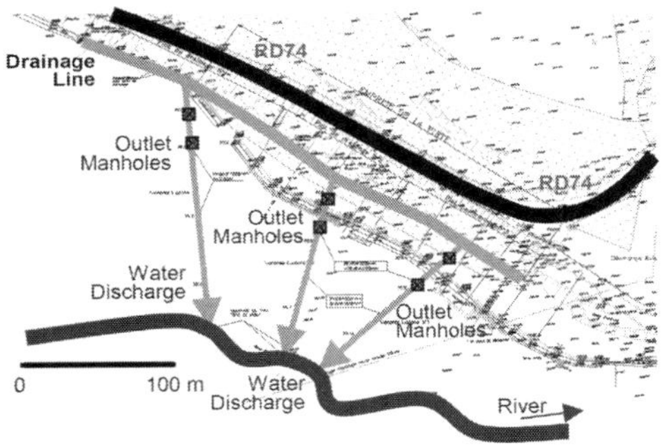

Figure 4: Scheme of the Siphon Drainage System at Goncourt.

as to protect the equipment of the Siphon Drainage technology against frost action and mechanical destruction, was excavated first. Protective drainage manholes made of concrete prefabricates Ø800 mm were placed at the position of each drain. The trench was partially backfilled by gravel and a perforated plastic duct was placed at the bottom to be used as a surface drain. The wells were drilled at Ø250 mm from the top of the manholes after backfilling the trench. The drains are 12.5 m deep and spaced at 5 m. The wells were equipped as standard opened piezometers. A perforated plastic pipe Ø110/114 mm was inserted into the boreholes and the space between the borehole and the screen was filled with filtrating gravel of 2–4 mm. An Air-lift to clean the wells was applied.

The collected groundwater was directed through siphon pipes (one for each drain) in three evacuation lines. The water collected by the drains is sucked out of the wells by the natural siphon principle through 10/12 mm diameter plastic siphon pipes inserted into each well and pulled through the burned ducting pipes between neighbouring manholes to three crosspoint manholes and then to six downstream placed common outlet manholes. Flushing systems for each drain were installed in the outlet manholes. The siphon tubes were primed by water under pressure and the outlet endings of the siphon pipes were connected to relevant flushing systems. A system of continual groundwater monitoring by piezometers was established.

After the system was put in function, the groundwater level in each well was lowered to 8–11 m below the terrain. In the initial stages of the drainage function, the global water discharge reached 250 l/min (3.5 l/min per drain). The current overall values of the water flow through the system are around 70 l/min (1 l/min per well). As of 2010, the system had already been in operation for seven years without any defects. Regular periodical maintenance of the drainage is carried out. No significant problems with the stability of the area occur anymore.

4 RD95 – Aigueblanche

On the national road RD95, for 3.680–3.800 km, considerable ground movements and problems with instability have been occurring for more than ten years. The instability was characterized by several active landslide zones developed on a steep slope in the altitudes between 700–850 m above sea level (Fig. 5). The rate of deformations varied from 10 to 70 mm per year and according to the inclinometry measurements, the depth of active shear planes varied between 6 and 14 m beneath the surface. The geological settings were designated as favourable for groundwater circulation and surface water infiltration. The ground is created by mixed rockmass and soils and it is composed of shales of variable degrees of weathering and clays. The initial piezometric levels were observed at a few levels between 1.5–3.0 m.

The results of a research study were that the sliding is too complex to stabilize the slope entirely by implementing mechanical barriers as pile walls or "Berlin" walls or some other method. The water table had been observed to be too high at the site. Geotechnical studies considered that the establishment of a drainage network with an 8 m depth efficiency should allow certain improvement of the safety factor, but still the drainage solution would not allow the complete stabilization of the slide. However, this solution was expected to slow down the movements and thereby reduce the deformations of the roadway. The gravity driven drainage system by means of Siphon Drains was chosen to achieve these goals.

Since the purpose of the drainage was just to reach a certain groundwater drawdown and the scale of the affected area was too large, exceptionally, the Siphon Drainage line was placed in the middle of the instable slope, below the road (Fig. 6). The drainage network was created by a single line of a total length of 130 m by 26 vertical Siphon Drains equipped with protective manholes (800 mm in diameter) with a spacing of 4 m. The drainage trench (for construction of the manholes, placing a collector at Ø200 mm for siphon tubes and a perforated collector at Ø150 mm for surface water) was up to 2.5 m deep (to realize the

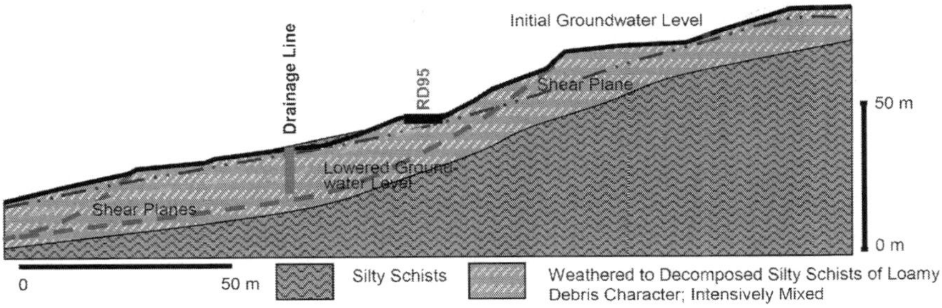

Figure 5: Cross-section of the affected slope at Aigueblanche.

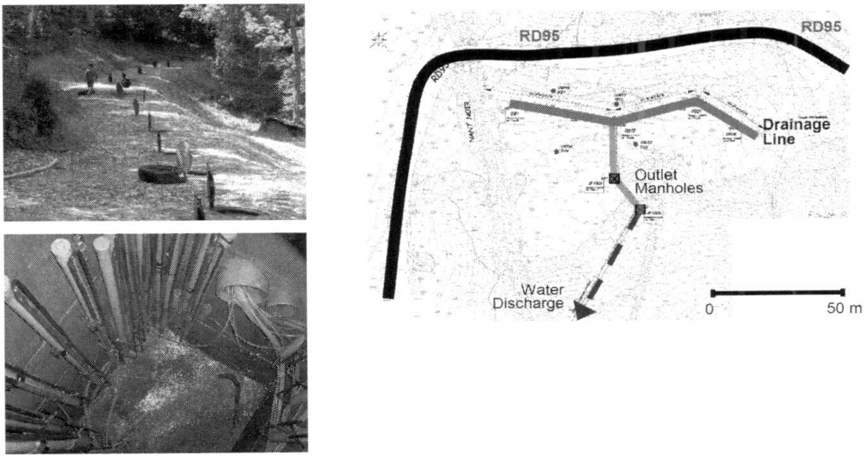

Figure 6: Scheme of the Siphon Drainage **S**ystem at Aigueblanche.

drawdown as deep down as possible). After construction and backfilling of the trench, the boreholes were drilled up to 15 m beneath the surface. Two outlet manholes were executed in the slope below the drainage line. The outlet manholes were constructed from prefabricated rectangle concrete units of a 1500 mm. The outlet manholes were connected to the drainage line by the same type of trench as the trench between the drains, including a Ø200 mm ducting pipe (collector) for siphon tubes and a Ø150 mm perforated collector for surface water. The drains were constructed and equipped as standard opened piezometers. The diameter of drilling was 250 mm. A perforated plastic pipe Ø110/114 mm was inserted into the boreholes and the space between the borehole wall and the screen was filled with gravel of grading 2–4 mm. The water collected by the drains is sucked out of the wells by the natural siphon principle through 10/12 mm diameter plastic siphon pipes inserted into each well. After the lower situated outlet manhole, the collected ground and surface water was directed by gravity downstream, to an opened concrete ditch. A system of continual groundwater monitoring by piezometers was established.

After the system was put in function, the groundwater level in each well was lowered to 10–12 m below the surface. The maximal registered flow for one single drain was 1.7 l/min. For reasons of extreme water inflows into some drains, the capacity was increased at these places by additional siphon tubes inserted into the wells or by changing the diameter of the tubes. By doing this, the maximal managed water flow was increased to 15 l/min. In 2010, the system had been in operation for four years without any defects. Regular periodical maintenance of the drainage is carried out. No problems with stability of the area occur anymore.

5 RD104 – Saint Priest

The road in the vicinity of Saint Priest was deformed due to general slope instability caused by a high water table (Fig. 7). A section of approximately 100 m of the state road RD104 was in danger if nothing was done.

The bedrock at the site is characterized by marls of different degrees of alteration and different degrees of compaction. The marls are alternating with limestones. It is though that the rockmass is slightly jointed. Quaternary deposits are developed as heterogenous debris, the thickness of which increases uphill from 0 m below the road embankment to 5 m in the slope above the road. This was caused by earthworks during the road construction. The upper groundwater level should have been lowered and was initially following the boundary between the debris and the marls. It was also observed to be very shallow in the marls below the road embankment. The upper water level was oscillating strongly according to precipitation. The lower groundwater level was explored in the marl formations, but it was thought not to have an influence on the stability problems.

In order to guarantee the water table lowering, the Siphon Drains method was selected as the most convenient for permanent water drawdown without electricity consumption. By using vertical wells, the water could have been easily lowered up to the depths required by the geotechnical designer in order to improve the general stability of the slope (Fig. 8).

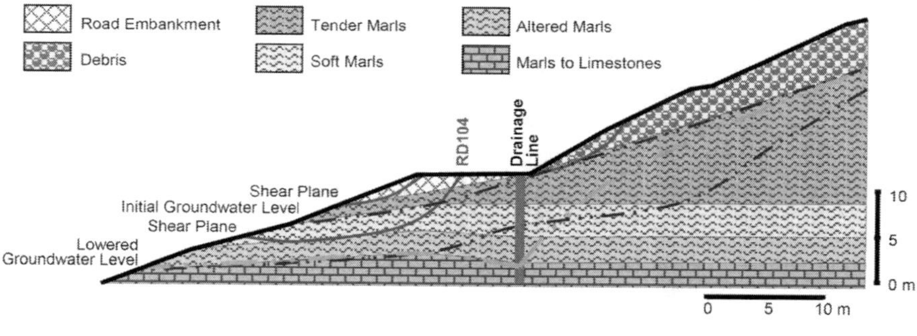

Figure 7: Cross-section of the affected slope at Saint Priest.

The drainage network was arranged in a single line containing 37 drains. The line was placed just next to the road. The standard arrangement of the Siphon Drainage system was kept at this site. A trench of 1.5 m minimum depth in which to place all necessary ducting, as well as to protect the equipment of the Siphon Drainage technology against frost action and mechanical destruction, was excavated. To realize the drawdown as deep down as possible, the trench was excavated up to 2 m depth. Protective drainage manholes made of concrete prefabricates Ø800 mm were placed at the position of each drain. The trench was completely backfilled with gravel. A perforated plastic duct was placed at the bottom to be used as a surface drain. The wells were drilled at Ø250 mm from the top of the manholes after backfilling the trench. The drains are 13.5 m deep. The distances between the drains are 3 m. The wells were equipped as standard opened piezometers. A perforated plastic pipe of Ø110/114 mm was inserted into the boreholes and the space between the borehole and the screen was filled with filtrating gravel of 2–4 mm. An air-lift to clean the wells was applied.

The collected groundwater was directed through siphon pipes (one for each drain) in two evacuation lines. The water collected by the drains is sucked out of the wells by the natural siphon principle through 10/12 mm diameter plastic siphon pipes inserted into each well and pulled through the burned ducting pipes between neighbouring manholes to two crosspoint manholes and then to four downstream placed common outlet manholes. Flushing systems for each drain were installed in the outlet manholes. The siphon tubes were primed by water

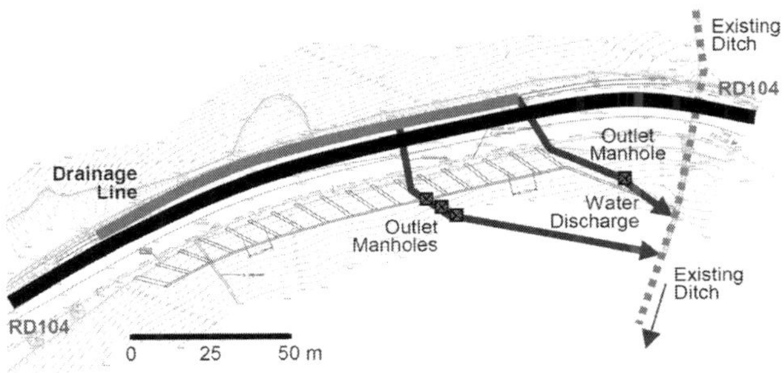

Figure 8: Scheme of the Siphon Drainage System at Saint Priest.

Figure 9: Deformations of the road and view at the Saint Priest site.

under pressure and the outlet endings of the siphon pipes were connected to relevant flushing systems. The final water evacuation was realized to an existent opened ditch.

A system of continual groundwater monitoring was established by means of 14 vibrating wire piezometers placed into the same wells. Telemetry was set up so the data could be downloaded, checked and processed anytime in the office and provided online to the client. After the system was put in function, the groundwater level in each well was lowered to 9–13 m below the surface. At the moment of preparing this article (soon after installation of the system), no information about water flows were known. Regular periodical maintenance of the drainage is carried out.

6 TGV Corridor – Chabrillan

The "Chabrillan" is a major cutting excavated in 1998 on the TGV high speed train link in France. It is located 530.300,20 km south of Valence. The cut is 1000 m long, with a maximum depth 35 m. In summer 2000, minor deformations on the access road were noticed. Later on, larger scale movements of a few centimetres in width resulted in a narrow fissure on the slope above the road. By the end of 2000, the fissure developed into a major feature of 30 m in length. It was suggested that movements and slope failures could affect the cutting and the train corridor. In 2001, the maximum recorded lateral ground movement reached about 1 mm/day towards the rail line and a total volume of 1.200 000 m^3 was in movement along two shear planes that were identified at the site.

The geology comprises molasse deposits principally formed by sandstones and fresh water limestones with the presence of karstic features and interlayers of plastic clayey marls (Fig. 10). Alpine tectonics is apparent by the presence of thrust planes inclined at 15–20° towards the cutting. Groundwater was observed at two levels, generally at depths of 20 m and 33 m below the crest of the cutting. Inclinometer records confirmed the presence of two main levels of ground movement at 19 m and 30 m. The total volume of the material in movement was indicated as 1.200.000 m^3. It was proposed to make an unloading cut and to excavate 600.000 m^3 of material in the slope behind the railway cutting to relieve the disturbing forces driving the slope instability. After performance of the excavation, the groundwater level was monitored in 2001–2005 and observed at 0–8 m beneath the base of the cut. In 2002, new movements were registered.

To control the groundwater level and to reduce the deformations, several dewatering schemes were considered. A 10 m deep trench for a length of 150 m was designed to be excavated from the base of the unloading cut. This solution suffered from a number of limitations: in particular, the trench might increase the risk of new shear failures. Subhorizontal wells were rejected due to considerable length and minimal efficiency due to generally low permeability and complex aquifers. To achieve the required drawdown, a deep drainage system was assumed as the most appropriate approach. Immerged pumps were not selected due to their poor efficiency at low permeabilities and inflows and for high energy requirements. The limitations of the gravity Siphon Drainage system lay in the

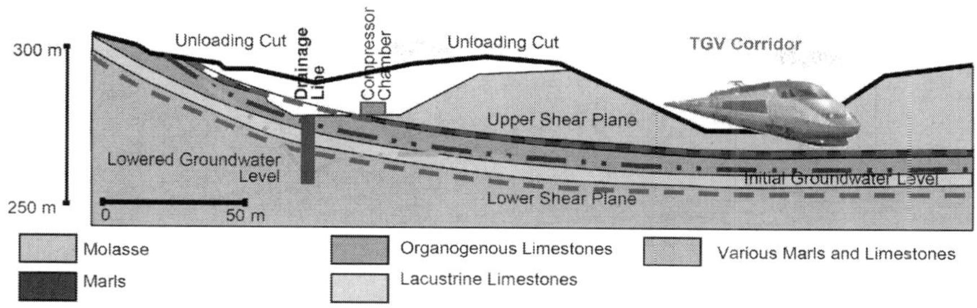

Figure 10: Cross-section of the affected slope at Chabrillan.

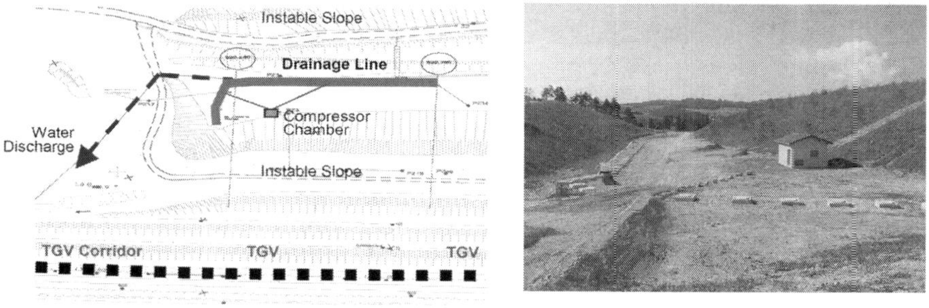

Figure 11: Scheme of the Electropneumatic Drainage System at Chabrillan.

plain gradient of the ground surface. Based on the slide characteristics, the site morphology, the required drawdown and the total construction, operation and service expenses of the drainage, a scheme by Electropneumatic Drains was selected as the most effective solution.

The design comprised 47 wells arranged in a 150 m long line (Fig. 11). The wells were bored at 200 mm diameter into 12–20 m depth and equipped with electropneumatic pumps. A slotted PVC casing of 103 mm internal diameter was installed to full depth. The annulus was filled with a fine graded gravel filter of 2–4 mm size. Each drain was equipped with a protective manhole of 1.5 m depth. All tubings and cables were led from wells to the compressor chamber where the compressors (2x30 kW, one as a back–up), the control panel and the accessories were installed. The chamber was designed as a simple brick house (4x6 m). A comprehensive system of instrumentation and monitoring was established. The monitoring comprised continual water level measuring, inclinometry and water flow observations. An alarms system by GSM was set-up for alerts in the case of high water level or any disconnection problems. Internet Explorer was chosen as the interface for the drainage system operation and monitoring results online checking.

The groundwater was lowered from the original 0–8 m to 11–15 m beneath the base of the unloading excavation. The drawdown and drainage efficiency comply with the requirements. The maximal total flow rates recorded reached 8–280 l/min (0.2–6 l/min per drain). The actual moderate flow rate alternates at around 0.2 l/min per well (8 l/min in total). In 2010, the system had been in operation for four years without any defects. Regular periodical maintenance ofthe drainage is carried out. No new slope deformations were registered after start-up of the system. In 2010, a special French price “Ivor” for quality and for innovative solution was given to Electropneumatic Drainage at this site.

7 Conclusions

Corridors of roads, highways or railways are frequently surrounded by slopes – natural hills or artificial cuts. The presence of groundwater in the slopes is one of the most negative factors that can affect the stability and functionality of the infrastructure tracks. Groundwater lowering is one of the ways to stabilize slope movements by improving the properties of soils and rocks and avoiding or reducing the risk of new deformations and damages of the structures. The method of groundwater lowering is very important. In this article, two innovative alternatives of dewatering by systems of deep vertical wells were introduced.

Since many of the stability problems of roads appear in non-urbanized areas without good access for a source of power, the gravity method of Siphon Drainage seems to be an ideal and very effective solution for long-term permanent groundwater lowering without any need for electricity for water pumping. In the case that the water table should be reduced by up to 8–12 m beneath the surface and the expected water inflows are around 2–15 l/min per well, this method provides a reliable solution.

The method of Electropneumatic Drains represents a flexible system of dewatering which does not require any specific morphology of the treated sites. It is suitable for urgent, emergency and even temporary solutions of deeper seated problems (up to 60 m depth). In the case that the groundwater level is supposed to be reduced in cycles (continuous pumping is not expected), the expenses for electricity and maintenance of the equipment can be saved by use of a single conventional compressor (10–50 kW/ 10–15 bars for 20–150 wells up to 60 m depth) instead of many submersible pumps of the same rated power input. A huge amount of water per well (up to 70 l/min) can be managed to be pumped.

References

[1] Bomont, S., Mrvík, O., Back Experience from Two Cases of Stabilization of Coastal Landslides by Innovative Deep Drainage Systems, *Proc. of the 11th Baltic Sea Geotechnical Conference*, Vol. 1, pp. 19–26, 2008.

[2] Bomont, S., Mrvík, O., Back Experience of Innovative Deep Drainage Systems for Slopes Dewatering and Landslides Stabilizations, *Geoinzynieria – drogi, mosty, tunele*, 04/2008 (19), pp. 60–65, ISSN 1895–0426, 2008.

[3] Mrvík, O., Bomont, S., Application of Innovative Method of Deep Drainage by Siphon Drains for Stabilization of Slopes of Former Opened–Cast Brown Coal Mine “Most – Lezaky” (Czech Republic), *Geotechnika,* 2/2009, pp. 20–25, ISSN 1211–913X, 2009.

Technical protection measures against natural hazards taken by the Austrian Federal Service for Torrent, Erosion and Avalanche Control

F. J. Riedl
Austrian Federal Service for Torrent, Erosion and Avalanche Control, District Office "Upper Inn Valley", Austria

Abstract

The focus of the following paper lies in the association between the theory of the planning part and the external practical work, demonstrated on a specific technical protection measure against the torrent "Farstrinne". Experiences during the building period in the year 2008 (damages on the technical mitigation measure caused by two flooding events) led to reconsider the measures undertaken until that time. The daily experience is absolutely necessary to work out protection measures against natural hazards (torrents, avalanches, landslides, rock fall) as well as for further planning works.

Initially, the Austrian Federal Service for Torrent, Erosion and Avalanche Control will be introduced and especially the District Office "Upper Inn Valley".

Keywords: Austrian Federal Service for Torrent, Erosion and Avalanche Control, technical protection measures, natural hazards, torrents, avalanches, rock fall, landslides, hazard zone mapping.

1 Introduction

1.1 Austrian Federal Service for Torrent, Erosion and Avalanche Control

The main tasks of the Austrian Federal Service for Torrent, Erosion and Avalanche Control are divided into four category groups:

- hazard zone planning (risk prevention),
- planning of technical mitigation measures against natural hazards,
- building construction of the protection measures and
- expertise for the public authorities.

WIT Transactions on Engineering Sciences, Vol 67, © 2010 WIT Press
www.witpress.com, ISSN 1743-3533 (on-line)
doi:10.2495/DEB100111

Therefore, the assignment of the Austrian Federal Service for Torrent, Erosion and Avalanche Control enfolded from risk assessment up to event management in case of extreme events caused by natural disasters.

RISK ANALYSIS
In the Risk Analysis the definition of the watershed areas, mentioned in the Forestry Law of 1975 [3], monitoring and survey activities in the catchment areas and analysing of hazards and risks are the main parts of our assignments. Furthermore, documenting extreme weather events, collection of the data and management of statistical databases are part of our risk analysis.

RISK ASSESSMENT
In the Risk Assessment the focus lies within the provision of basic decision-making processes. Therefore the hazard zone planning requested in § 11 Forestry Law [3] also maintains the creation of guidelines and regional planning studies. The provision of planning principles and standards are part of the daily work of the Austrian Federal Service for Torrent, Erosion and Avalanche Control.

RISK MANAGEMENT
After extreme weather events and the treatment of permanent settlements by torrents, avalanches, landslides or rock falls the detailed planning and realisation of protection measures against these natural hazards is essential. In collaboration with the University of Natural Resources and Applied Life Sciences in Vienna, passive protection measures (monitoring and early warning systems) are also be used and installed in specific torrents [8].

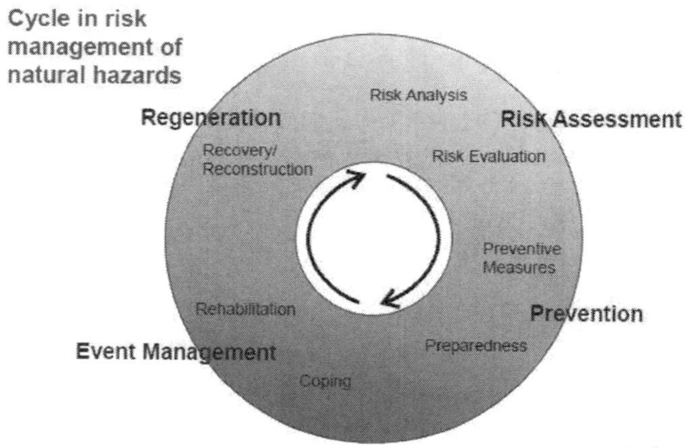

Figure 1: Cycle of risk management after Rickenmann [5].

1.2 District Office "Upper Inn Valley"

The District Office "Upper Inn Valley", situated in Imst, is responsible for the torrent, avalanche, rock fall and erosion protection in the two mountainous districts Landeck and Imst in the western part of Tyrol. The total area covers 3,320km^2 and the population, scattered in 54 communities, reaches about 96,446 people. The most important source of income of the population of these two districts is tourism. There are annually around 11.1 million overnight stays, 69% of them being in the winter.

In the Imst and Landeck districts there are about 340 catchment areas (torrents), including 664 harmful avalanche tracks endangering the permanent settlement of space, which is about 7.4% of the total. Taking the population on the permanent settlement of space, there is a high population density of 391 persons per km^2. The described high risk and hazard potential leads to the fact that the District Office "Upper Inn Valley" is one of the most endangered regions of the Alps affected by natural hazards.

The main focus of action lies within the protection against avalanches. Thus, in the last 50 years, extensive construction works (snow supporting structures) were erected. In total about 130km snow supporting structures and 500ha afforestations near the timberline reduce the risk for the permanent settlements.

The District Office "Upper Inn Valley" invests about 9.3 million € annually in protective measures, especially in the construction and planning works [8].

61%	to protect against avalanches
24%	protection against torrents
8%	in afforestation near the timberline
7%	to protect against rock falls and landslides

Due to permafrost and the melting of the ice cores in the soil caused by global warming, the rock fall activities have almost doubled in the last 10 years.

2 Technical protection measure torrent: "Farstrinne", community Umhausen, district Imst

2.1 General catchment characteristic of the torrent "Farstrinne"

The catchment area of the hydrological western drained torrent "Farstrinne" is about 5.5 km^2. The highest elevation point in the catchment area is 3010m above sea level, which means a difference in height of nearly 2060m. The upper catchment area is very steep and characterized by numerous gullies. The middle part is mostly rocky with a slope topography of 17-35%. The striking debris cone of the Farstrinne was built up by numerous ancient events, where many blocks of about 4m^3 were deposited. The slope topography of the cone itself is about 20% and nowadays most of the ground level is used for agricultural work.

From the hydrological point of view, the catchment area belongs to the Inner-Alpine, Continental, Climate Zone. The annual precipitation is approximately

710 mm and the previously measured maximal rainfall was 96.1 mm in 24 hours on 4 Nov. 1966.

The maximum runoff was calculated by the software program ZEMOKOST [4] by monitoring a repeat period of time of 150 years. ZEMOKOST is a recalculated program contemplating the surface runoff und the surface roughness after Markart et al. [2]. For this existing torrent the maximum runoff HQ_{150} is about 50m³ per second and the expected bedload deposit considering this maximum runoff, is about 80.000 m³. The bedload deposit was calculated by the empiric formula after Hampel regarding the inclination of the debris cone topography of 20%.

2.2 Historical information of the torrent “Farstrinne”

The impressive historical information of damages, caused by the torrent “Farstrinne”, leads to the permanent settlement “Östen” situated on the topographical right side of the debris cone. On the one hand this historical information was gained from records taken by the population itself together with the literature of Fliri [1] on the other hand. The smaller events with a repeat period of time of 2-3 years are not mentioned below. These events affect the major road into the Ötztal Valley and leads particularly to an embargo of the main route.

15 July 1749	80 buildings of the permanent settlement “Östen” were destroyed; 9 persons died.
1760	a debris flow destroyed 20 buildings; no one was injured.
4 July 1762	8 buildings of “Östen” were buried.

In the years 1767, 1791, 1807, 1830. 1854, 1855, 1857, 1862, 1874, 1878 and 1891 several debris and hyperconcentrated flows destroyed the church, buildings and agricultural areas of “Östen”. The temporary stone construction made by the population itself was completely destroyed and there was no safety anymore.

7 May 1991	A debris flow accounted for about 40.000m³ manifested itself above and on the major road into the Ötztal Valley as well as between the two temporary protection dams. The torrent broke out on the topographical right side where about 2ha agricultural areas were affected and destroyed.

In July and August 2008 two events happened during the construction work of the technical mitigation measures. These two events divided into a smaller (July) and a larger one (August) amounted for a total volume of deposit of about 25.000m³. The protection measures were partly destroyed and the most of the material went down to the downstream deposition basin above the major road. The downstream basin with a capacity of about 15.000m³ was too small and therefore the major road was closed because of the overwhelming material.

2.3 Idea behind the planning and construction works

Caused by the events up to the year 1991 the District Office "Upper Inn Valley" initiated several technical protection measures on the torrent "Farstrinne". On the upper part of the debris cone a deposition basin with a capacity of about 80.000m³ was supposed to prevent large destruction. The discharge unit is characterized by a ground fissure of about 2.0m. The idea behind this construction was that small and middle events are transferred downstream to the already existing second deposition basin. This downstream deposition basin will be enlarged to enable the capacity to reach up to 50.000m³. The third step in securing the permanent settlement is to expand the existing longitudinal dams where big armourstones should protect the dams against erosion processes.

In the building period of 2008 the main task was to build and complete the upper deposition basin with a deposit capacity of 80.000m³. The technical details are as follows: 1300m³ concrete, 60 tonnes steel, 60.000m³ bulk material and 23.000 tonnes building blocks. The concrete-steel phased discharge unit was planned and realized with an angular steel girder against transported timber during an event, fig. 4. This sorting should ensure a normal runoff of a hyperconcentrated water or debris flow.

The angular steel girder was dimensioned on the base of a 1.5 times hydrostatic normal water pressure. Several ancient deposition basins with a similar discharge unit were constructed and dimensioned in the same way the years before.

Figure 2: The upper deposition basin with the discharge unit and total deposition capacity up to 80.000m³.

2.4 Hazard zone mapping

In Austria, hazard zone mapping is mentioned in the Forestry Law of 1975 [2] and in regulations according to the decree of the Ministry for Agriculture and Forestry of 1976. In this decree there are guidelines for hazard mapping and directives for delineation of hazard zones. The delineations are based on specific criteria for torrents and avalanches, fig. 3. The delineations for torrents are based on flow or erosion depths and energy lines of the water. For avalanches, the pressure of the avalanches is essential considering a repeat period of time of 150 years. In the red hazard zone, the installation or construction of buildings is not advised and not allowed. In the yellow hazard zone the construction of buildings are possible and acceptable under certain conditions.

In the present case study of the torrent "Farstrinne" there are about 8 buildings situated in the red hazard zone and 38 buildings in the yellow hazard zone.

Austria: criteria for delimitation of hazard zones

Kind of Hazard	Zone	Frequent event (10y)	Design event (150y)
Avalanche	Red	p > 10 kN/m², d > 1.5 m	p > 10 kN/m², d > 1.5 m
	Yellow	p > 1 kN/m², d > 0.2 m	p > 1 kN/m², d > 0.2 m
Fall processes	Brown	Whole area affected by the process	
Stagnant water (ponding)	Red	HQ1 > 0.2 m; HQ10 > 0.5 m	d > 1.5 m
	Yellow	HQ1 < 0.2 m; HQ10 < 0.5 m	d < 1.5 m
Water flood	Red	energy line > 0.25 m	energy line > 1.5 m
	Yellow	energy line < 0.25 m	energy line < 1.5 m
Gulley erosion	Red	erosion possible	d > 1.5 m
	Yellow	see water flood	d < 1.5 m
Sediment deposition	Red	deposition possible	d > 0.7 m
	Yellow	see water flood	d < 0.7 m
Bank erosion	Red	-	Stable slope criteria
	Yellow	-	safety buffer
Debris flow	Red	-	Border of debris flow lobes
	Yellow	-	-
Management	Blue	Area reserved for technical/biologic counter measures	
Management	Magenta	Conservation of the actual state of the area	

Figure 3: Delineations for the hazard zone mapping in Austria after Rickenmann [6].

3 Damage, experiences and knowledge

3.1 Steel construction and damage on the discharge unit

As mentioned before, in the year 2008 two events affected the technical protection measure during the construction work. The second one partly

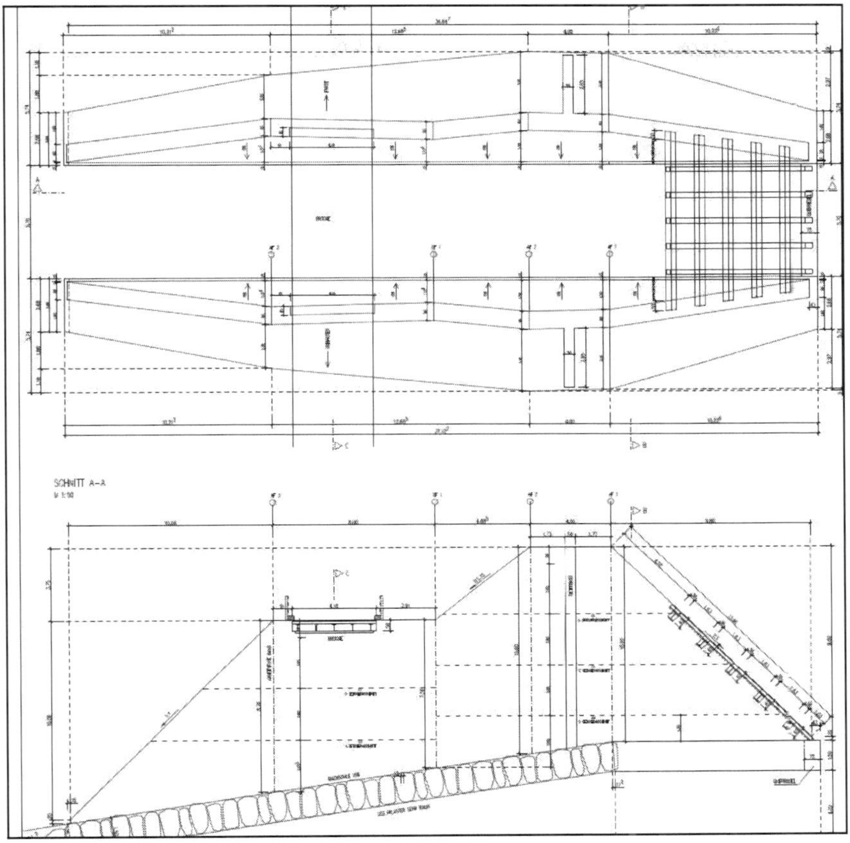

Figure 4: The angular steel girder of the upper discharge unit.

destroyed the discharge unit and completely damaged the angular steel girder, fig. 5. The major road into the Ötztal Valley was closed for two days but nevertheless no casualties were registered. This was the first time that such an event destroyed the steel girder that had been in normal use.

After this event the learning outcome was that the normally used dimensions for such steel girder on the basis of 1.5 times hydrostatic normal water pressure are not able to provide the occurring density and energy of a hyperconcentrated or of a debris flow. In this particular case, a large block crashed into the middle of the steel girder and an inflection of the steel occurred, fig. 5. This led to the collapse of the deposited material.

3.2 Types of events in the year 2008

In the year 2008 the smaller as well as the middle event were hyperconcentrated up to slurry flow after Selby [7]. The quantity of water was extremely high and the block sizes, which were transported in the medium, reached up to $7m^3$ means

Figure 5: Deformation of the steel girder after the second event 2008.

Figure 6: Hyperconcentrated up to slurry flow and deposition in the downstream basin after the second event 2008.

about 14 tonnes. In general, the events with a repeat period of time of 2-4 years are hyperconcentrated up to slurry flows. Only the large events, in comparison to the years 1749 or 1760 are typical debris flows and also the ancient debris cone indicates such former events.

3.3 Conclusions

Regarding the planning phase of such technical mitigation measures it is important to underline that natural disasters have a certain residual risk. Nevertheless, basic experiences which are gained during the period of time should inflow into further planning works. The main conclusion based on this illustration is that the 1.5 times hydrostatic normal water pressure is not sufficient enough to resist against hyperconcentrated or debris flows. The wingspan of the steel construction up to 6.0m was also too extensive and smaller wingspans should be preferred.

Another point of view is the cost and height of the residual risk. If the residual risk is higher it is also possible to redevelop certain damages on the technical protection measures appearance during a flood event. Using a stronger steel-profile caused by a bigger dimension leads to enormously higher costs. Due to the higher costs of a stronger steel-profile, several redevelopments can also be undertaken. Therefore, the costs (monetary value) versus the benefits (residual risk) should be conscientiously regarded by the planners.

4 Costs of the construction works undertaken on the torrent: "Farstrinne"

The total costs of these technical protection measures are about 3.0 Mio €. The upper deposition basin costs 1.6 Mio €, the longitudinal dams costs 0.3 Mio € and enlargement of the smaller basin downstream costs about 1.1 Mio €. As a requirement for the projects undertaken by the Austrian Federal Service for Torrent, Erosion and Avalanche Control, it is obligatory that in every project a cost/benefit analysis is done. This is the basis for the financial federal support. In the case of the "Farstrinne" the cost/benefit analysis represented a quotient of 1.4 for the benefit. It is not mandatorily necessary to reach a value above 1.0 but it is an important index and important for the decision-making process.

References

[1] Fliri, F., *Naturchronik von Tirol: Tirol - Oberpinzgau - Vorarlberg – Trentino*, Beiträge zur Klimatographie von Tirol: Austria, 1998.

[2] Markart, G., Kohl, B., Sotier, B., Schauer, T., Bunza, G., Stern, R., *Provisorische Geländeanleitung zur Abschätzung des Oberflächenabflussbeiwertes auf alpinen Boden-/Vegetationseinheiten bei konvektiven Starkregen*, BFW-Praxisinformation 3/2004: Innsbruck, 2004.

[3] Ministry for Agriculture and Forestry (BMLFUW), *Forestry Law 1975*, BGBl. Nr. 440/1975, idF BGBl. Nr. I108/2001: Austria, pp. 3–5, 43–45, 1975.

[4] Kohl, B., Stepanek, L., *ZEMOKOST - neues Programm für die Abschätzung von Hochwasserabflüssen*, BFW-Praxisinformation 8/2005: Innsbruck, pp.21–22, 2005.
[5] Rickenmann, D., *Risk Analysis of Mountain Hazards*, Lecture Notes of the University of Natural Resources and Applied Life Sciences in Vienna, part A Introduction: Vienna, pp. 29, 2006.
[6] Rickenmann, D., *Hazard and Risk Assessment*, Lecture Notes of the University of Natural Resources and Applied Life Sciences in Vienna, A1 Hazard Mapping part I (A): Vienna, 2006.
[7] Selby, M.J., *Hillslope Materials and Processes*, Oxford University Press: Oxford, pp. 299–304, 1993.
[8] http://www.die-wildbach.at/

Section 3
Risk assessment and hazard mitigation

The distribution of debris flows and debris flow hazards in southeast China

F. Wei[1,2], Y. Jiang[2], Y. Zhao[2], A. Xu[2] & J. S. Gardner[3]
[1]*Key Laboratory of Mountain Hazards and Surface Process, Chinese Academy of Sciences, China*
[2]*Institute of Mountain Hazards and Environment, Chinese Academy of Sciences, China*
[3]*Natural Resources Institute, Clayton Riddell Faculty of Environment, Earth and Resources, University of Manitoba, Canada*

Abstract

Debris flows commonly occur throughout southern China, posing a hazard to people, property and infrastructure. Although debris flows occur more frequently and with greater magnitude in mountainous southwest China, where they have been studied extensively, their impacts on society are greater in southeast China, where population densities and land use intensities are greater. A number of recent disastrous debris flow events in southeast China have drawn the attention of researchers and the government. Using historical records and field investigations, this paper describes debris flow distribution, casual and contributing factors, hazards and disaster prevention and mitigation measures in southeast China.

Keywords: debris flows, distribution, hazard, southeast China.

1 Introduction

The west to east topographic profile of China is characterized by three general units and declining elevation. At higher elevation in the west is the Qinghai-Tibet Plateau, followed by the Yunnan-Guizhou Plateau, Sichuan Basin and Loess Plateau, and finally the lower elevation region of middle and eastern China. Debris flow occurrence is most common in the higher elevation areas of the first two units and the transition zone between them. However, a number of disastrous debris flow events in the third unit, particularly its southeastern part,

WIT Transactions on Engineering Sciences, Vol 67, © 2010 WIT Press
www.witpress.com, ISSN 1743-3533 (on-line)
doi:10.2495/DEB100121

have drawn attention to the debris flow hazard there. Although the distribution density, occurrence probability and scale of debris flow here are less than in southwest China, when they do occur debris flows often result in significant casualties and property loss For example, several debris flows triggered by Typhoon Rananim in the mountainous area in the north of Zhejiang Province on 12 Aug. 2004 produced 42 fatalities and 46.88 million CNY direct economic losses [1]; a debris flow induced by heavy rainfall (208 mm in 4 hours) in Zhangdun, Jianyang county, Fujian Province on 21 May 1989 caused 47 fatalities, 394 injuries and 46.35 million CNY direct economic loss [2]. As a result of these and other debris flow disasters, more attention is being paid to debris flow hazard research in southeast China. He [3] analyzed the process characteristics, causal mechanisms and principal controlling factors of the main geological disaster sites in Fujian Province, and proposed hazard and disaster mitigation measures ranging from management to technological applications. Wang [4] described the environmental context and current status of geological disasters in Zhejiang Province, and recommended technological and management prevention and mitigation measures. By analyzing flood and torrent hazard conditions and features in Guangdong Province, Lin [5] pointed out problems in the local government's work in organizing flood hazard mitigation and prevention and presented a strategic plan for such work. These researchers analyzed the geological disasters in three provinces of southeast China and offered a framework for hazard and disaster mitigation. However, as yet there is no research or countermeasures intended specifically for debris flow hazard. Using data obtained by field investigation and historical records, this paper examines the distribution of debris flows and describes their characteristics, and discusses countermeasures for debris flow hazard and disaster mitigation in Southeast China.

2 Characteristics of the study region

The area under consideration in Southeast China includes three provinces, Zhejiang, Fujian and Guangdong, and is located between 109°27′~122°42′E and 20°14′~30°54′N and contains a total area of 429,200 km^2 (Fig. 1) .

2.1 Topography

The study area lies in coastal area of Southeast China and is part of the third and lowest topographic tier of China (Figure 1). Low-relief hills are widely distributed in the region and plains areas are distributed along the coast. The altitude is less than 500 m in most of the and the highest elevation is 2,158 m (Huanggangshan, peak of Wuyi Mountain range) (Figure 2). The low-relief hills taking up more than two thirds of the region consist mainly of the Minzhe hills in Fujian and Zhejiang and the Lingnan hills in the north of Guangdong and the south of Fujian. Influenced by intense crustal movement in the Stage of Yanshan, faults, folds and magmatic intrusions and lava extrusions are common in Minzhe Hills. Topographically, it features small intermountane basins and

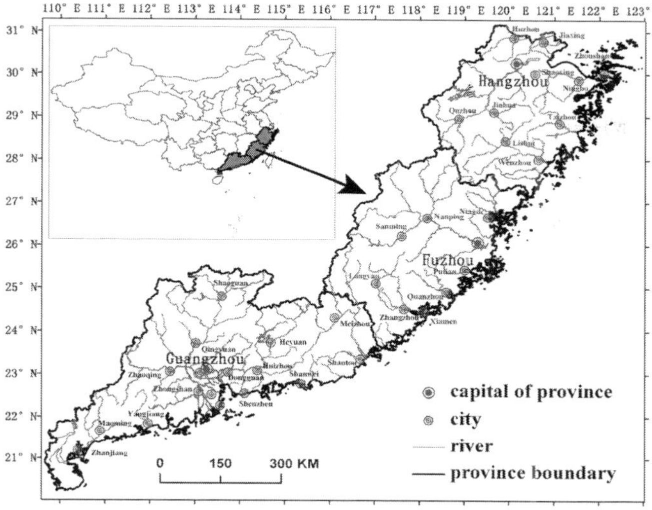

Figure 1: Study region and its position.

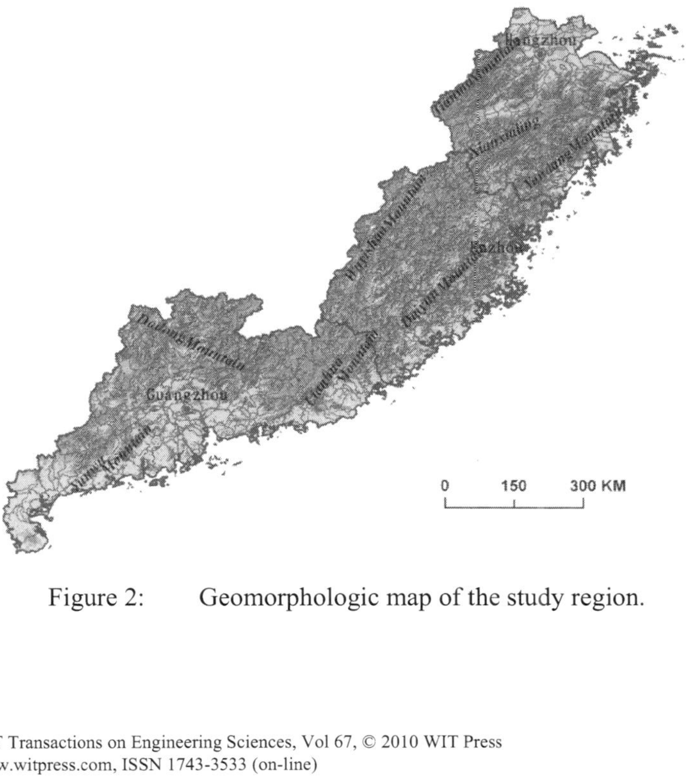

Figure 2: Geomorphologic map of the study region.

plains lying at the base of the widespread hills. Two ranges paralleling the coast line compose the relief frame of the Minzhe Hills. The westerly range, with average altitude of more than 1,000 m, consists of the Wuyishan Mountains, the Xianxialing Mountains and the Tianmu Mountains. The hills are composed mainly of ancient metamorphic rocks and Paleozoic sedimentary strata. The eastern range, with an average altitude about 800 masl, consists of the Lianhua Mountains, Daiyun Mountains and Yandang Mountains, stretching from southeast to northwest. The mountains are composed mainly of igneous rock and granite. The Lingnan Hills mostly are granite hills with round appearance and strong spherical weathering.

2.2 Geology

Tectonic activity has been frequent in the study region as it is located at the intersections of the Pacific Plate, Eurasian Plate and the Philippine Plate. The crustal structure is complex and the topography and geomorphology are strongly influenced by it [6]. The region has experienced several strong tectonic movements. The influence of the Caledon episode is strongest among these movements with the fold mountain paralleling the coast being formed in the Caledonian, with further folding during the Indosinian [7]. Active faults are strongly represented in Fujian and Guangdong. These fault zones generally consist of several faults paralleling the coast in a northeast direction. Some bead-shaped basins developed along the edges of these faults (Figure 3) [8].

Massive magmatic intrusion and volcano activity during the Mesozoic produced extensive areas of medium-acidic intrusive rocks and volcanic rocks. Thus, volcanic rocks cover most of Zhejiang and eastern Fujian. Red sandstones

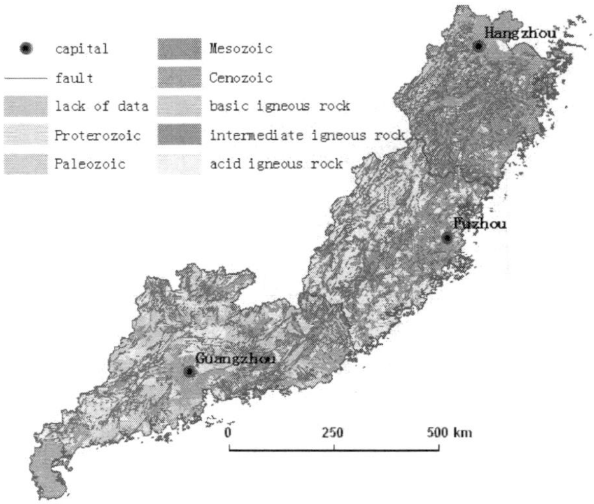

Figure 3: Geological map of the study region.

and limestones of Cenozoic age are found in intra-montane basins and valleys. The coastal plain and river valleys are covered mainly with Quaternary sediments.

2.3 Climate

Southeast China is located in the East Asian monsoon region with a subtropical climate with the exception of southern Guangdong which is characterized by a tropical climate. The average annual temperature decreases from north to south with. Zhejiang in the north has an average annual temperature of 15-18°C, Fujian in the middle has 20.1°C and Guangdong in the south has 19-24°C. Average annual precipitation is high and increases from north to south and from the coast inland to the mountains. The average annual precipitation at Zhejiang is 980-2000 mm, at Fujian is 1,452 mm and at Guangdong is 1,300-2,500 mm. The precipitation is distributed unequally during the year with more than 85% occurring during April to October. Frontal and cyclonic systems generate precipitation with two annual peak periods, one in May and June and the other in August and September. The study region is affected by typhoon very seriously. Most typhoons influencing China land from this region and bring extreme rainfalls, which is the main type of rainfall that induces debris flows.

3 Debris flow distribution and its principle

3.1 Debris flow types

Debris flows in Southeast China may be divided into two types: slope debris flows and gully or valley debris flows. Much of southeast China is made up of

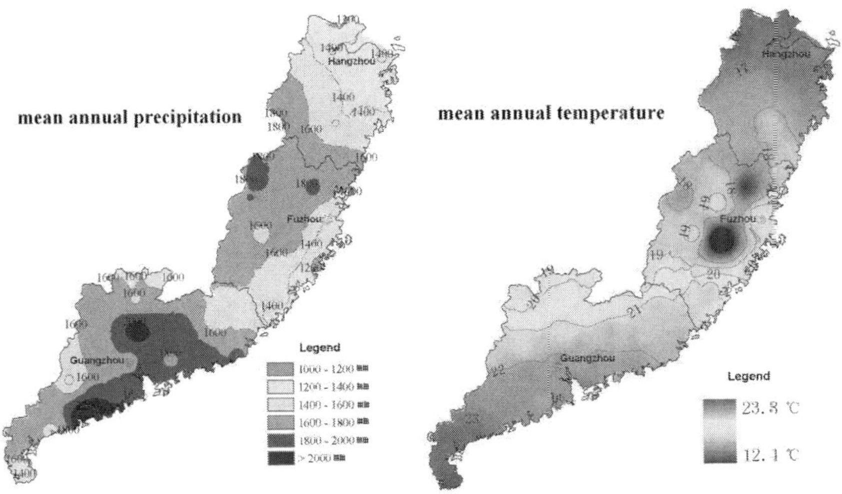

Figure 4: Distribution of mean annual precipitation and temperature.

low mountains and hills with small altitude difference, making the debris flow tracks relatively short. The main debris flow type is a slope debris flow, meaning that it is initiated and flows on an open slope. Damage done by slope debris flows is usually relatively minor because of the short running distance and low velocity. A second type of debris flow, the gully or valley debris flow originates in a small catchment and follows the path of the main catchment outflow. This type of debris flow has strong damage potential because of long movement distance and high velocity. Although gully or valley debris flows are rare in the study region, most serious debris flow disasters are induced by them. The debris flows discussed in this paper are mainly of the gully or valley type.

The human activities in southeast China are the most intense in China. Wastes produced by road building and mining may accumulate and induce debris flows. For example, the debris flow event of 15 June 2002 in Dalu village, Minqing County of Fujian Province was caused by accumulated waste debris from an expressway construction in the channel of the Dalu valley. This debris flow destroyed a factory built on the fan in the valley and cased more than 10 million CNY in direct economic loss.

3.2 Debris flow distribution

There are 463 known debris flow valleys in Southeast China (Figure 5). 291 debris flow valleys are in Zhejiang Province, distributed in 46 counties; 107 valleys are in Fujian Province, distributed in 38 counties; and 65 valleys are in Guangdong Province, distributed in 24 counties. In Zhejiang Province, debris flows are distributed mainly in the mountains of western and southern Zhejiang,

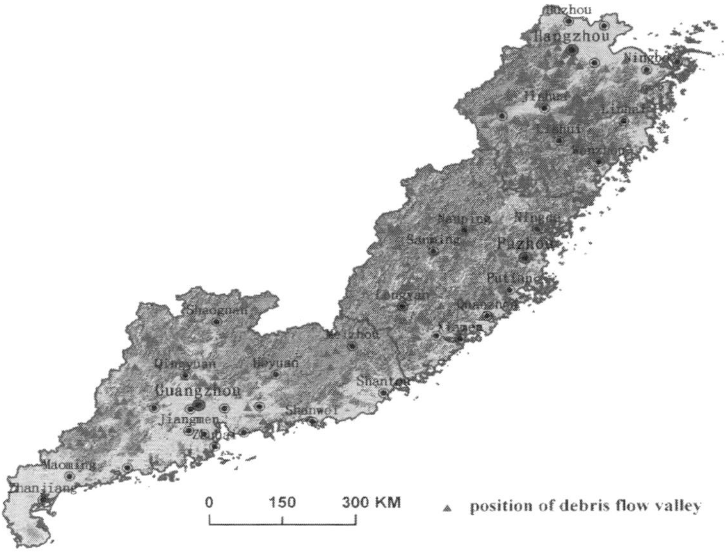

Figure 5: Distribution of known debris flow valleys.

including Wenzhou, Lishui, Quzhou and Jinhua. In Fujian Province, debris flows are distributed in all mountainous and hilly areas except the coastal area. In Guangdong Province, debris flows are distributed mainly in mountains of northern Guangdong.

In terms of topographic units, Southeast China's debris flows are mainly distributed in the Minzhe Hills and Nanling. In the Minzhe Hills, debris flows are mainly distributed in Yandang-Dalao Mountains, Kuocang-Donggong-Daiyun-Bopingling Mountains and Longmen-Xianxia-Wuyi mountains. All of these mountains are oriented parallel to the coast line. In Nanling, debris flows are mainly distributed in Yao Mountains and Jiulian mountains.

3.2.1 Topographic characteristics

In order to analyze the debris flow distribution by topography, the study region was divided into grids with the size of 3km×3km. The distribution of known debris flow valleys in the grids with different elevation differences is analyzed and then the probability of debris flow valleys appearing in the grids with different elevation differences is evaluated as in Fig. 6. According to Fig. 6, the probability increases along with a gain in elevation difference. And, the probability increases dramatically after elevation difference reaches 600 m. This result is similar to what has been found in Southwest China but with but some differences. The probability would decrease with gain in when it reaches a specific threshold number in southwest China [9], while it is monotonically increasing in Southeast China cases. This is mainly because Southwest China mostly consists of mid and high mountains with steep slope and large elevation difference, while Southeast China mostly consists of low mountains and hills with small elevation difference.

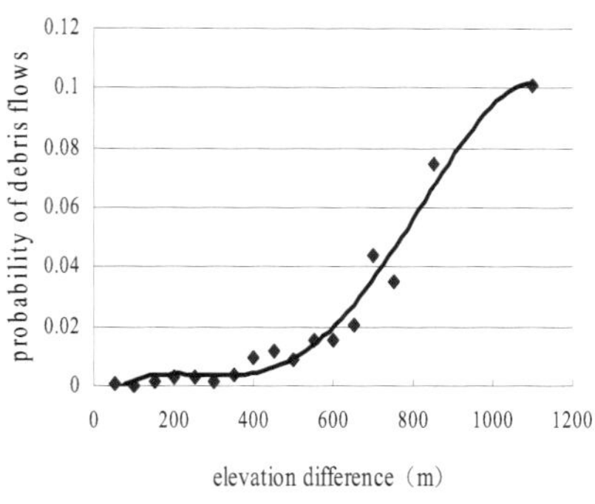

Figure 6: Distribution of debris flow in topography.

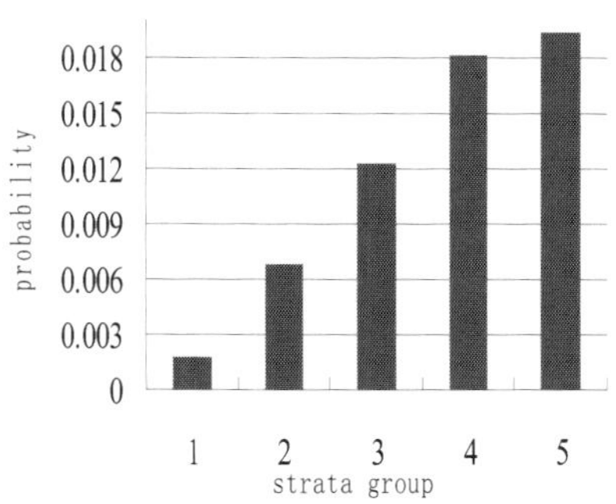

Figure 7: Distribution of debris flow by geological strata.

3.2.2 Debris flow distribution by geological strata

In order to analyze debris flow distribution by strata type, the complex strata are classified as five groups (Table 1). This classification is based on the reference 9 and revised according to the research of Zhang et al. [10]. The distribution of known debris flow valleys in five stratum group is analyzed and then the probability of debris flow valleys appearing in each group is evaluated as in Fig. 7. According to Fig. 7, the probability has evident differences in the five groups.

3.2.3 Debris flow distribution in relation to faults

In order to analyze debris flow distribution in relation to faults, the fault density in every grid is calculated with GIS tools according to the fault length in each grid and the area of the grid, and then the probability of the known debris flow valleys appearing in the grids with different fault density is evaluated as in Fig. 8. According to Figure 8, the probability of debris flow valleys increases with an increase of fault density.

3.2.4 Debris flow distribution in time

Influenced by the monsoon climate, debris flows in Southeast China are mainly distributed in summer (from May to October), and more concentrated during

Table 1: Classification of strata.

Category	lithology		
	Sedimentary rock	Magmatic rock	Metamorphic rock
1	Dolomite, charcoal grey and coarse lamellar limestone, concreted, siliceous limestone, cherty limestone	thick-bedded rhyolite, thick-bedded andesite, basalt	Quartz, quartz vein, diabase, diabase vein
2	Quartzy sandstone, siliceous conglomerate, bleached limestone, quartzy siltstone	Fine and medium-grain granite, diorite, gabbro, andesite, tuff, rhyolite porphyry, basic igneous rock, ultrobasic rock, alkaline granite, diabase, porphyrite	Marble, quartz schist, amphibolite, serpentine, metamorphic basalt
3	Sandstone, marlite, sandy and siliceous mudstone, conglomerate	Volcaniclastic rock, porphyritic coarse-grained granite, syenite	Schist, slate, granulite, metamorphic liparite, metasandstone, gneissose
4	Shale, siltstone, coal-bearing strata, semi-consolidated rock, unconsolidated sandstone	Volcanic debris	Phyllite
5	Quaternary loose deposit (loess, alluvial deposition, diluvial deposition, slope wash and moraine), mild clay, clay, clay sand, peat, mudstone		

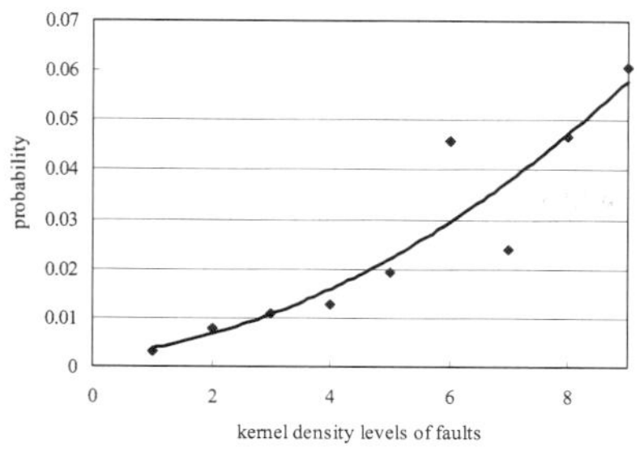

Figure 8: Distribution of debris flow in relation to faults.

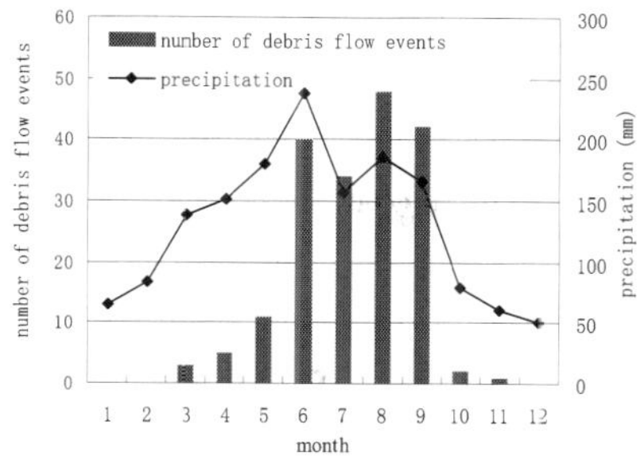

Figure 9: Annual distribution of debris events and precipitation.

June to September. The distribution of debris flow events in the most recent 30 years and average precipitation month by month in southeast China are illustrated in Fig. 9.The distribution of debris flows events in a year takes on a bimodal distribution coinciding with the distribution of precipitation. The first peak appears in June coinciding with the onset of the monsoon, while the other peak in August which is coincident with the typhoon season which often brings very heavy rainfalls.

4 Countermeasures for debris flow disaster mitigation

Because of the dense population and highly developed industries and infrastructure in southeast China, debris flow disaster mitigation is very important. According to the characteristics of debris flows noted above, the following countermeasures for debris flow disaster mitigation are proposed.

(1) Identification of debris flow valleys

Debris flow valleys and potential debris flow valleys are often ignored or misjudged in southeast China because debris flow frequency of occurrence is relatively low in most locations , vegetation conditions are good and the traces of past debris flows are not obvious as they are modified by fluvial processes and hidden by vegetation. Almost all of the serious debris flow disasters have occurred in these kinds of valleys. So distinguishing debris flow valleys and potential debris flow valleys is a very important initial countermeasure for debris flow disaster mitigation. The valleys which will be used for buildings and infrastructure must be completely assessed before and construction takes place.

(2) Preventing debris flow disasters by control works

Southeast China has a high population density, and the land resource is very valuable. Though a valley has been identified as a debris flow site, in most instances, the area's land use will not be given up. Thus, debris flow control works seem rather important. Debris flow control works must be constructed in the valleys where debris flows have occurred or have the potential for occurring.

In addition, the large magnitude of debris flows induced by strong typhoons, as compared with water floods, has to be considered in the design of debris flow control works.

(3) Intercepting floating woods

Because the economy of southeast China is highly urbanized and industrialized, most slopes in the hill and mountain areas have returned to forestland from cropland and the percentage of forest cover is high. As a result, many of the valley debris flows carry large numbers of logs and other vegetative material that adds to the flow magnitude and causes flow blockages at channel constrictions and at bridges and culverts. To prevent this, the logs should be intercepted before they move to control works, bridges and culverts.

(4) Debris flow prediction and warning systems

Most of typhoons approach the study region from the east and southeast and they are observable long before they make landfall, thus making forecasting and warning possible. Typhoons often bring heavy rainfall and trigger hazardous debris flow events. The magnitude of debris flows induced by typhoon rainfall is often very high and the peak discharge is often greater than the design standard of the control works. So debris flow prediction, forecasting and warning are important not only for debris flow valleys without any control works but also for those with control works. Debris flow prediction, forecasting and warning are active measures which may reduce deaths and injuries due to debris flows.

5 Conclusion

The number and frequency of debris flows is less in the low relief terrain of southeast China than in high relief southwest China where debris flow have been more commonly studied. However, because of the well-developed economy with its dense population and abundant infrastructure, debris flows pose a more serious risk and often lead to serious disasters in southeast China. Slope debris flows are the most common form in southeast China, but most of the serious disasters are caused by valley debris flows. The rapid development of the region has both exposed more people and property to the hazard and in some cases has contributed to the occurrence of debris flows. Based on the present data, 463 known debris flow valleys are distributed in three provinces of southeast China, but mainly in the Minzhe Hills and the eastern part of the Lingnan Hills, and more occur in Zhejiang and Fujian Provinces than in Guangdong. The debris flow distribution has obvious regularities in topography, stratum, fault and time. The probability of debris flows distribution is advanced with the growths of elevation difference and faults density. The distribution of debris flows events in a year takes on a bimodal distribution coinciding with the prime rain season and the typhoon season. Based on the debris flow characteristics in Southeast China, the best countermeasures for disaster prevention and mitigation are identification of potential debris flow valleys, prevention of debris flows by control works and flow regulation structures, interception of logs and other vegetative material and the use of debris flow prediction and warning systems.

Acknowledgements

This research is supported by the Research Fund for Commonweal Trades (Meteorology) (Grant number: GYHY200706037), the Key Project in the National Science & Technology Pillar Program (Grant number: 2008BAK50B04) and Chinese Academy of Sciences Visiting Professorship for Senior International Scientists (Grant No. : 2009Z2-13).

References

[1] Yuan, L., Cui, X., Wang, Z., et al. Cause mechanism of Xianrentan debris flow in Yueqing City, Zhejiang Province, *Journal of Natural Disasters*, 18(2), pp.150-154, 2009.(in Chinese)

[2] Ruan, F., Fang, D., Zhang, Q., et al. Then cause formation and characteristics of the debris flow event on May 21 in Longan, Jianyang. *Fujian Soil and Water Conservation*, (2), pp.48-52, 1989. (in Chinese)

[3] He, Y. Characteristics and mechanism of major geological hazards in Fujian Province and Protection and controlling method against them. *Geology of Fujian*, (4), pp.263-271, 1995. (in Chinese)

[4] Wang, J. Current situation of geological disasters and the prevention measures in Zhejiang Province. *Journal of Catastrophology*, 16(4), pp.63-66, 2001. (in Chinese)

[5] Lin, J. How to promote the prevention of mountain torrents disasters in Guangdong Province. *Guangdong Water resources and Hydropower*, (4), pp.84-85, 2004. (in Chinese)

[6] Zhao, P., Wang, J., Wang, J-A., et al. Characteristics of heat production distribution in SE China. *ACTA Petrologica Sinica*, 11(3), pp.292-305, 1995.

[7] Ren, J. The Indosinan orogeny and its significance in the tectonic evolution of China. *Bulletin of the Chinese Academy of Geological Sciences*, (9), pp.31-42, 1984. (in Chinese)

[8] Hu, H., Shen Y. Principal characteristics of vertical crustal deformation in Southeast China. *Seismology and Geology*, 12(2), pp. 121-130, 1990.

[9] Wei, FQ., Gao, KC., Hu, KH., et al. Relationships between debris flows and earth surface factors in Southwest China. *ENVIRONMENTAL GEOLOGY*, 55(3), pp.619-627, 2008.

[10] Zhang, J., Wei, FQ., Yu, SJ., et al. Susceptibility analysis of debris flow to rocks based on hydrological characteristics of their weathering products. *Chinese Journal of Rock Mechanics and Engineering*, 27(11), pp.2227-2233, 2008. (in Chinese)

Evaluation of sediment yield from valley slopes: a case study

F. Ballio, D. Brambilla, E. Giorgetti, L. Longoni,
M. Papini & A. Radice
Politecnico di Milano, Dept. I.I.A.R., Milan, Italy

Abstract

Hydro-geological hazards in alpine areas is a really common problem Many calamitous phenomena (such as debris flows, landslides, and others) are related to the sediment yield from the slopes of the valleys. Sediment yields are far from being fully understood and predictable, due to a lack of knowledge of the physical mechanisms underlying these processes and to the variability of the peculiar geomorphologic characteristics of river basins. Key unknowns are the medium- and long-term average sediment production, the recharge time of the sediment sources (and consequently the frequency of the yields), the triggering factors and the thresholds for activation. The manuscript documents the results of the estimation of sediment production for the basin of the Tartano valley in northern Italy. The basin is characterized by a significant presence of weak rocks (cataclastic, mylonitic), that makes considerable amounts of loose sediments available. In this work, semi-quantitative models were applied to evaluate the basin-scale, yearly sediment yield. Estimates sediment volumes were compared to records of sediment volumes extracted from an artificial reservoir located at the downstream section of the catchments. In addition, the spatial distribution of the sediment instability level was obtained, highlighting a significant heterogeneity of the river basin. Therefore, the relevance of the basin-scale modelling of sediment yields for off-site and on-site processes was discussed. The dependency of the sediment yield regime on the spatial and temporal scale supporting the evaluations was analyzed and discussed.

Keywords: river basin, sediment yield, scale-based evaluation, soil erosion.

WIT Transactions on Engineering Sciences, Vol 67, © 2010 WIT Press
www.witpress.com, ISSN 1743-3533 (on-line)
doi:10.2495/DEB100131

1 Introduction

Sediment transport is a key aspect of the life of river basins. Sediments are eroded from the valley slopes by exogenous and endogenous agents. Then, as nicely pointed out by Phillips [1], the common sense is that any sediment particle may be deposited on the same slope from which it has been eroded, or it can be involved in landslides, debris flows, or also reach a river stream to be conveyed downstream by the flowing water. Proper evaluation of the sediment yield is important from a technical point of view, for example to evaluate the tendency of the basin system to some undesirable conditions such as riverbed aggradation, reservoir sedimentation, as well as debris flows and landslides.

The sediment yield in a river basin results from the composition of a number of effects. The conceptual picture proposed by De Vente and Poesen [2] involves several types of sediments source (splash erosion, sheet erosion, rill erosion, gully erosion, bank erosion and mass movements) and some sink terms (depression, parcel, footslope and floodplain storage) whose combination determines the total sediment yield at a certain downstream section. According to Wasson [3], the separate modelling of all the processes for a final composition of the effects is hardly possible. On the contrary, a sediment yield modelling at a large scale is more feasible and, therefore, most desirable in order to obtain practical results.

The models for the estimation of the sediment yield fall within few categories, namely: the physically based models, the conceptual models, the empirical models and the semi-quantitative models (see, for example, the review by De Vente and Poesen [2]). In principle, the physically based models enable quantitative evaluations to be made, even though they require extensive data for a proper application. The other models are progressively simpler to use and provide semi-quantitative results. In addition, other models can be used for a relative evaluation of the tendency of the basin to instability, without a numerical output for the sediment productivity.

A crucial aspect of the evaluation of the sediment yield is the scale with reference to which the model is made. The reviews of De Vente and Poesen [2] and Wasson [3] span several orders of scales, from the small basin to the global scale. A significant relationship emerges between the spatial scale of analysis and the type of model that is most suitable: physically based models can be used for small parts of the basins, in which only few source or sink terms are present, and then that a hard composition is not required; semi-quantitative models can be used for basin-scale evaluations; finally, if a regional or global scale is considered, a sort of self-similar behaviour emerges and the sediment yield can be obtained by some universal-like equations depending only on the basin area.

This manuscript presents an analysis of the sediment yield for the Tartano basin, which is located in the Italian Alps. Given the above considerations, the evaluation of the sediment yield will be mostly conducted at a basin scale. The suitability of the obtained results for analysis of off-site and on-site processes is discussed. An evaluation of the response of yield estimation to the spatial and temporal scale of modelling is presented at the end of the manuscript.

2 The case study: description of the Tartano basin

The present study refers to the Tartano Valley, located in Northern Italy. The Tartano basin (which is 49 km^2 wide) is part of the Adda catchments; furthermore, the Tartano flows into the Adda along the upper course of the latter (that is, upstream of the Adda flowing into the Como lake). Elevation in the Tartano catchments ranges from 950 to 2250 m above sea level. The climate is defined as Alpine continental. Meteorological records show that the local temperature is subjected to strong altitudinal gradients in temperature and precipitation. The strong rainfall, low temperature, snow precipitation and high annual and day-time thermal range favour the activity of the morphogenetic processes related to erosion. Therefore, soil erosion is pronounced in the Tartano basin, as in all the upland Adda catchments.

An aerial picture of the basin is shown in Figure 1, together with the catchments boundary and the main hydrographic network. The Tartano River originates from two main tributaries, namely the Val Lunga and the Val Corta. The downstream section of the basin in Figure 1 is not placed where the Tartano merges with the Adda but at the Campo dam (located a few km upstream of the confluence), because the annual data on the reservoir silting at this dam shall be used for comparison with the estimations of the sediment yields described in the following.

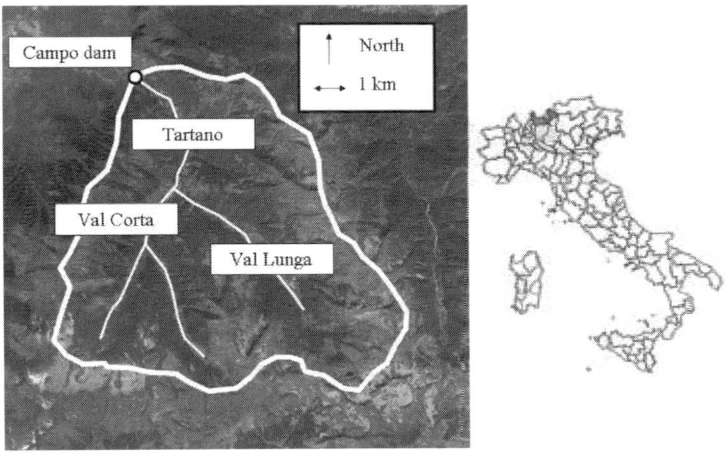

Figure 1: Aerial map of the river basin with indication of the main streams.

The vegetation of the basin is dominated by a forest of mountain pine (70 %). Different sediment sources occur in this valley: alluvial and colluvial storage, glacial deposit and colluvial breccia formed next to faults. Thus, this hydrographic basin is an effective prototypal case for sediment yield estimation. Structurally, the area under investigation belongs to the crystalline base of the Southern Alps, where the Gneiss of Morbegno emerges. The rocky substrate involves two systems of faults: NE-SW and NW-SE trending (Figure 2).

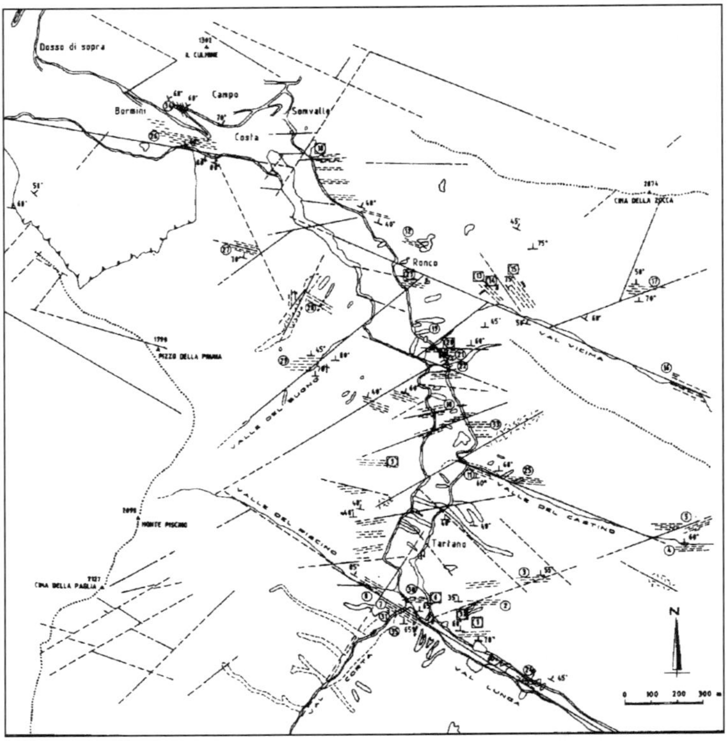

Figure 2: Fault lineaments in Tartano Valley.

Knowledge of the fault network is relevant for erosion processes: the colluvial breccias is due to the presence of weak rocks (band of fault rocks) surrounding fault lineaments. A great accumulation of material can be observed along faults and, during strong meteorological events, this mass can move rapidly along the slopes, feeding the solid transport of Tartano River. Erosion process and consequently sediment yield are very common in these fault rock bands, due to low geotechnical parameters and high degree of fracturation.

The yearly records of the sediment volumes taken out of the Campo dam are presented in Table 1. A mean annual sediment volume of 38038 m^3 is estimated.

Table 1: Annual sediment yield (SY) into the reservoir at the Campo dam.

Year	1991	1992	1993	1994	1995	1996	1997	1998
SY (m^3)	34073	43504	53605	36737	26264	39749	35314	32800
Year	1999	2000	2001	2002	2003	2004	2005	2006
SY (m^3)	41876	57299	43187	42022	22957	50083	21287	27844
Mean SY value (m^3)			**38038**					

3 Basin-scale evaluation of the annual sediment yield

This section describes the estimation of the annual sediment yield using several models: Gavrilovic [4], USLE (Wischmeier and Smith [5]) and RUSLE (Renard *et al.* [6]). All models were applied to the entire catchments closed at the Campo dam. This choice was most suitable for comparison of the estimated yields with the field data mentioned above.

The Gavrilovic model involves a semi-quantitative analysis for erosion estimation in a defined closed loop of the hydro-geological basin. This method was originally developed for catchments in the south of Yugoslavia. The basic concept of the model is that the sediment volume transported by the stream (G, m^3/year) depends on the sediment yield by soil erosion (W, m^3/year) and the sediment deposition in the watershed (through a sediment retention coefficient R,), according to the following equation:

$$G = W \cdot R \tag{1}$$

The calculation of the sediment yield W involves empirical coefficients (erodibility coefficient, soil protection coefficient, and erosion coefficient) and some physical characteristics (annual precipitation, temperature, average slope, and surface area):

$$W = T \cdot H \cdot \pi \cdot Z^{\frac{2}{3}} \cdot F$$

$$R = \frac{(l + l_i) \cdot \sqrt{O \cdot D}}{(l + 10) \cdot F}$$

$$T = \sqrt{0.1 + \frac{t}{10}}$$

$$Z = \Xi \cdot \Pi \cdot (\Phi + \sqrt{I}) \qquad \text{(2a, 2b, 2c, 2d)}$$

where: T is a coefficient of temperature, H is the mean annual rainfall (mm), F is the area of the watershed (km^2), Z is the coefficient of relative erosion, O is the perimeter of watershed (km), D is the mean difference in elevation of watershed (km), t is the mean annual temperature of the whatershed (°C), I is the mean slope of the watershed, l is length of the principal waterway and l_i is total length of the secondary waterways (km). The coefficient of relative erosion Z depends on several factors related to the soil and to the basin: Ξ (coefficient of soil cover), Π (coefficient of soil resistance to erosion) and Φ (coefficient of the observed erosion process). The values for Ξ, Π and Φ are chosen based on qualitative descriptions of the basin, to which some numerical ranges correspond. The present choice was: $\Xi = 0.2$ (coniferous forest with little grove, scarce bushes, bushy prairie); $\Pi = 1.6$ (Sediments, moraines, clay and other rock with little resistance); $\Phi = 0.8$ (50-80 % of the catchments area affected by surface erosion and landslides). All the parameter values chosen for the case study are described in Table 2, together with the yield computation results. For this simulation the worst meteorological condition was considered.

The coefficients Ξ, Π, Φ are crucial for model application because, as seen, only some range of values are suggested based on qualitative descriptions of the

Table 2: Annual sediment yield obtained from the Gavrilovic model.

T (° C)	H (mm/year)	I	l (km)	l_i (km)	F (km^2)	D (km)
3	1376	0.58	11.26	149.84	47.0	1.79
O (km)	Ξ	Π	Φ	W (m^3/year)	R	**G (m^3/year)**
29.22	0.2	1.6	0.8	45371	1.67	**52931**

basin. A sensitivity analysis was performed for these coefficients, even though the coefficient values were changed remaining within the range proposed for the qualitative features chosen. Table 3 shows the influence of these parameters on the final result. This sensitivity analysis provides G values ranging between 34380 and 63160 m^3/year. The mean value of sediment yield obtained by field surveys (Table 1) is near to the lower limit of the range while the upper limit can explain sediment yield in the worst years.

Table 3: Sensitivity analysis for the coefficient in the Gavrilovic model.

	Base case	Case 1	Case 2	Case 3
Ξ	0.2	0.15	0.2	0.2
Π	1.6	1.6	1.8	1.6
Φ	0.8	0.8	0.8	0.85
Z	0,50	0,37	0,56	0,52
W (m^3/year)	45371	29470	54139	47568
G (m^3/year)	52931	34380	63160	55494
G variation (%)		-35	+19	+5

USLE (Wischmeier and Smith [5]) is another empirical model used for sediment budget definition. This method was devised in the 1950s by the USA Department of Agriculture and evaluates the annual soil loss in farmland neglecting sediment deposition.

$$E = R \cdot K \cdot L \cdot S \cdot C \cdot P \tag{3}$$

where: E is the average annual erosion, R is the rainfall-runoff erosivity, K is soil erodibility, L is the slope length, S is slope gradient, C is crop cover and management factor and P is support/conservation practices factor.

R and K are two dimensional parameters that represent synthetically the aggressiveness of erosive agent (R) and soil characteristics (K), while L, S, C, P are dimensionless factors.

USLE model was revised and a new method (called RUSLE, Renard *et al.* [6]) was presented. The key difference with respect to USLE model is that in RUSLE the factors L and S are combined into a single factor LS. R is mathematically defined as the product between total kinetic energy in a single meteoric event and the maximum intensity in a period of 30 minutes during the same event. The sum of every erosive event during one year provides the annual value; the mean of annual values extended to a pluriannual period provides the value of R factor. The K factor explains the intrinsic aptitude to erosion of the soil. In USLE and RUSLE application the problem is related to the choice of K parameter because the equation used for quantification of K was defined through some experimental analysis conducted on different geological conditions.

Results obtained from the application of USLE and RUSLE models to the Tartano basin are listed in Table 4.

Table 4: Sediment yield estimated with USLE and RUSLE.

E_{USLE} (m^3/year)	24708
E_{RUSLE} (m^3/year)	12587

Some comments can be made with reference to the presented computations. The Gavrilovic model overestimates the mean annual sediment yield into the reservoir at the Campo dam, while USLE and RUSLE underestimate it. This contrasts with a reasonable expectation of USLE providing larger values (it shall be remembered here that this model neglects sediment deposition). It should be borne in mind, however, that the Gavrilovic model was calibrated with reference to basins presenting significant similarities with the Tartano catchments, while USLE and USLE-derived methods were devised for rural basins in the USA. Despite the variability of the results, all the models correctly estimate the order of magnitude of the yield.

4 Discussion: limitations of the basin-scale modelling

The evaluation of the sediment yield documented above is representative of the global, average behaviour of the river basin. On one hand, fluctuations of the annual sediment yield can be observed in the records previously shown (Table 1), indicating a long-period variability. On the other hand, as pointed out, for example, by De Vente and Poesen [2], the different parts of the river basin may contribute very differently to the average sediment yield and the largest volumes of sediments may come from small definable areas. Indeed, the conceptual picture by De Vente and Poesen [2] according to which the dominant sediment source and sink terms vary with the basin dimension holds also for the different homogeneous areas within river catchments.

The spatial variability of the tendency to the sediment yield can be visualized by means of thematic maps obtained from the application of specific models as SHALSTAB (Dietrich and Montgomery [7]) or USPED (Moore and Burch [8]). The former is a physically based model for shallow landslides, where the stability analysis of a slope is combined with the rainfall regime. For SHALSTAB a digital elevation model of the case study is necessary (in this work, a DEM with 1:10.000 scale was used). It is possible to insert the instable known areas in case a back analysis has to be conducted by the software. The model considers the ratio between effective rainfall and soil transmissivity (q/T): areas with lower values of this ratio are the more instable. So SHALSTAB evaluates for each DEM cell the stability ratio q/T and provides as output a grid of logarithmic values classified in seven intervals from "chronic instability" to "stable". USPED is a simple model which predicts the spatial distribution of erosion and deposition rates for a steady state overland flow with uniform rainfall excess conditions for transport capacity limited case of erosion process. The rates of erosion and deposition depend on the variation of transport capacity in the considered domain. Where transport capacity increases erosion takes place while where it decreases water releases sediments causing deposition. The results obtained are depicted in Figures 3 and 4. The spatial variability of the basin is evident. In addition, a lot of instable areas are present. These instable parts are localized on the steep slopes for ShalStab. For USPED, it is possible to see that the instabilities roughly correspond to the hydrographic network.

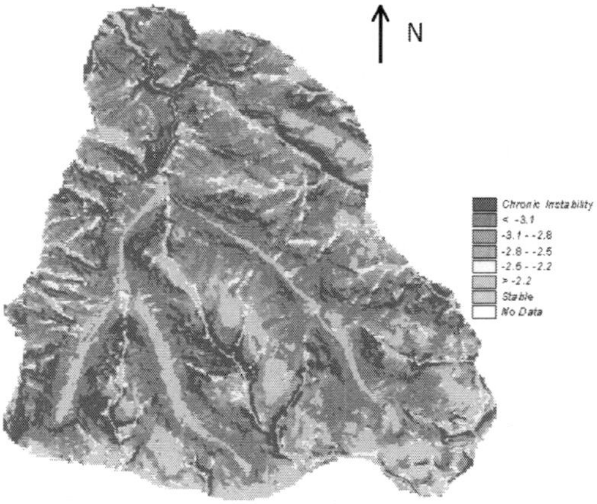

Figure 3: Application of SHALSTAB model.

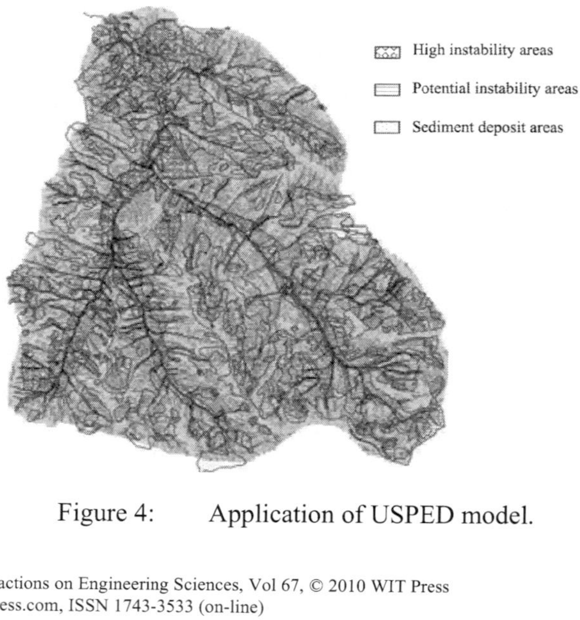

Figure 4: Application of USPED model.

The presented maps indeed support the concept that the expected sediment yield present significant spatial variability. As argued by De Vente and Poesen [2], an evaluation of the sediment yield at a basin scale is suitable for the analysis of off-site process (for example, the silting regime of a reservoir situated downstream of the last section of the basin, as that providing the field data used here). By contrast, there are several on-site processes that are conditioned by local sediment yields: among others, reference is made here to sediment transport within the water courses and debris flows along the valley slopes. A proper modelling of such processes requires adequate boundary condition in terms of the sediment yield. For the case of river sediment transport, Mandelli *et al.* [9] have identified three major flaws of the models based on lumped variables, namely: (i) the spatial scale, already discussed; (ii) the temporal scale, since the majority of the sediment volume conveyed by rivers is transported in the few days with largest discharge within the year whilst the yield modelling provides only an integral value for the whole year; (iii) the granulometry of the yielded sediments, which is a key piece of information for all the sediment transport models (e.g., Chanson [10]) but is not furnished by the models for sediment yield. Similar considerations may hold for debris flow phenomena (e.g., Iverson [11]). All the above considerations stimulate local-scale modelling for the sediment yield within short periods, for example those where significant events take place. An attempt of such modelling for a parcel within the Tartano basin is presented in the next section.

5 Scale issues in the evaluation of the sediment yield

This section presents some preliminary attempts to evaluate the response of the sediment yield evaluation to the spatial and temporal support scale of the modelling. For the evaluation of the spatial scale effect, the USLE and RUSLE models were applied to some sub-basins of the catchments. The chosen sub-basins are depicted in Figure 5 and correspond to: the entire Val Corta basin (see section 2); the entire Val Lunga basin; a pasture-covered parcel within the Val Lunga (henceforth indicated as subL); a wood-covered parcel within the Val Corta (sub C). For the temporal scale effect, the event-induced sediment yield was estimated using the MUSLE model. The latter was proposed by Williams and Berndt [12] for the evaluation of the sediment loss during a single rainfall event (Y_S). The proposed equation is:

$$Y_s = R_d \cdot K \cdot LS \cdot C \cdot P \tag{4}$$

where Y_s is the sediment yield (tons per storm) and R_d is a runoff factor, while the other symbols have the same meaning as in previous USLE and RUSLE models. For the application of the MUSLE model two events were considered, with return period of 10 and 100 years, respectively. Results of the evaluation are displayed in Table 5.

The relative variability of the results using the Gavrilovic, USLE and RUSLE models for the Tartano catchments was already discussed above (section 3). Now, the results for the different sub-basins and USLE models can be taken, for

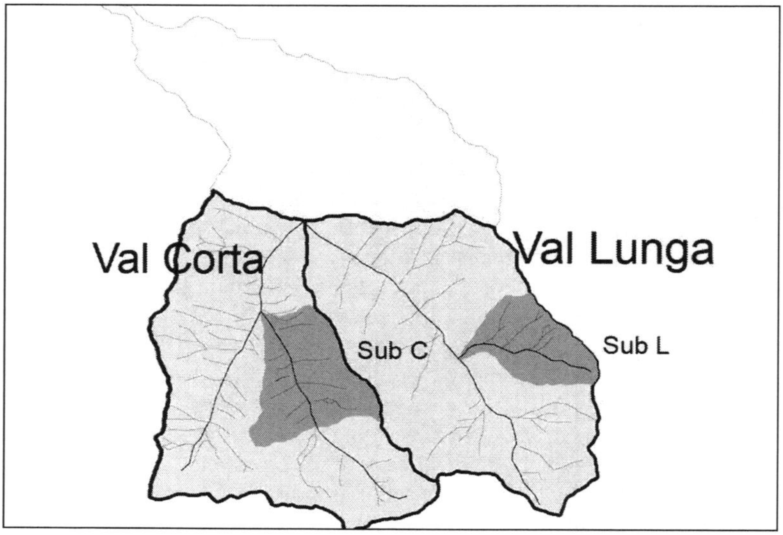

Figure 5: Sub-basins considered in the Tartano catchments.

example, to discuss the spatial scale effect. It appears that the spatial scale has no significant effect as long as the considered sub-basins are large enough to ensure the presence of several types of surface (Val Corta and Val Lunga sub-basins). By contrast, as parcels with only one type of soil cover are considered (sub L and sub C) a dramatic effect of the spatial scale appears, which is due to the presence of few types of surface (in other words, moving to little scales terrain features become predominant). The effect of the temporal scale is even more pronounced: considering events with significant intensity, huge sediment yields are obtained compared to the yearly ones (it should indeed borne in mind that the low number in Table 5 for the event-induced yields refer to very small durations compared to a whole year). In addition, the previously mentioned effects of the sub-basin surface are detected also for events with a short duration.

Table 5: Estimated scale response of sediment yield.

Basin	Tartano	Val Corta	Val Lunga	Sub L	Sub C
Area (km^2)	49	18	18	2.3	3.1
	Annual specific sediment yield (tons/ha/year)				
Gavrilovic	22.5				
USLE	10.5	10.8	12.0	44.7	12.1
RUSLE	5.4	7.5	8.7	28.5	8.3
	MUSLE evaluation of event-induced sediment yield (tons/ha)				
10-year event	0.7	0.9	1.1	6.3	0.8
100-year event	1.0	1.2	1.3	8.3	1.0

6 Conclusions

This manuscript considered the evaluation of the sediment yield in a mountain basin by means of semi-empirical models, with particular reference to the test case of the Tartano Valley in northern Italy. The estimation of the sediment yield was performed at the basin scale using the Gavrilovic, USLE and RUSLE models. The results obtained with these models presented a significant variability, yet in all cases the order of magnitude of the annual sediment yield was consistent with that obtained from periodic observation of sediment volumes extracted from a reservoir located at the downstream section of the basin.

The application of stability models like SHALSTAB and USPED provides significant pieces of information about the spatial heterogeneity of the basin in terms of the surface features and of the consequent tendency to soil erosion. The internal dynamics of the basin is visualized showing erosion and deposition areas. The scaling issues in sediment yield processes were discussed in light of this variability, which is expected to influence also the spatial distribution of the specific sediment yield. It was indeed found that the spatial scale of modelling influences the expected values of the specific sediment yield when small parcels having homogeneous soil cover are considered. In addition, the temporal scale of modelling was considered, showing that short-duration events with significant return period lead to concentrated (in time) sediment yields which may be dangerous even if the total amount of yielded sediments is low compared to the yearly one.

In the authors' opinion, applying models with reference to a variety of spatial and temporal scales might enable synoptic analyses of the basin dynamics to be made. Much further work is however needed to achieve a comprehensive perspective on these issues.

Acknowledgement

Participation of the third author to this research has been possible thanks to the funding by Regione Lombardia within the Project "Risk by Sediment Sources in Mountain Environments" (RISSME).

References

[1] Phillips, J.D., Fluvial sediment budgets in the North Carolina Piedmont, Geomorphology, **4**, pp. 231-241, 1991.

[2] De Vente, J. & Poesen, J., Predicting soil erosion and sediment yield at the basin scale: Scale issues and semi-quantitative models, *Earth-Science Reviews*, **71**, pp. 95-125, 2005.

[3] Wasson, R.J., What approach to the modelling of catchments scale erosion and sediment transport should be adopted?, *Modelling erosion, sediment transport and sediment yield*, eds. W. Summer & D.E. Walling, Technical Documents in Hydrology, n. 60, UNESCO, Paris, pp. 1-11, 2002.

[4] Gavrilovic, S., Bujicni tokovi i erozija, Gradevinski kalendar, Beograd, Serbia, 1976.
[5] Wischmeier, W.H. & Smith, D.D., Predicting rainfall erosion losses, Agric. handb. 537, USDA, Agricultural Research Service, Washington, DC, 1978.
[6] Renard, K.G., Foster, G.R., Weesies, G.A., McCool, D.K. & Yoder, D.C. (coordinators), Predicting Soil Erosion by Water: A Guide to Conservation Planning with the Revised Universal Soil Loss Equation (RUSLE), Agric. handb. 703, USDA, Agricultural Research Service, Washington, DC, 1997.
[7] Dietrich, W.E. & Montgomery, D.R., SHALSTAB: a digital terrain model for mapping shallow landslide potential, available online at the link http://calm.geo.berkeley.edu/geomorph/shalstab/index.htm, 1998.
[8] Moore, I.D. & Burch, G.J., Modeling erosion and deposition: topographic effects, *Transactions of ASAE*, **29**, pp. 1624-1640, 1986.
[9] Mandelli, M., Longoni, L., Papini, M., Roncoroni, F. & Radice, A., Modellazione del trasporto di sedimenti sul bacino del Tartano (Valtellina), *GEAM*, **XLVI(2)**, pp. 53-64, 2009.
[10] Chanson, H., *The hydraulics of open-channel flow: an introduction*, Elsevier Butterworth-Heinemann, 1999.
[11] Iverson, R.M., The physics of debris flows, *Reviews of Geophysics*, **35(3)**, pp. 245-296, 1997.
[12] Williams J.R. & Berndt H.D., Sediment Yield production with the universal equation using runoff Energy factor, *Present and Prospective Technology for Predicting Sediment Yield and Sources*, USDA, ARS-S-40, pp. 244-252, 1975.

Shallow landslide full-scale experiments in combination with testing of a flexible barrier

L. Bugnion[1] & C. Wendeler[2]
[1]*Swiss Federal Institute for Forest, Snow and Landscape Research WSL, Switzerland*
[2]*Geobrugg AG, Switzerland*

Abstract

Open shallow landslides occur regularly in a wide range of natural terrains. Generally, they are difficult to predict and result in damages to properties and disruption of transportation systems.

In order to improve the knowledge about the physical process itself and to develop new protection measures, full-scale experiments were conducted in Veltheim, Switzerland. Material was released down a test slope into a flexible barrier. The flow, as well as the impact into the barrier, was monitored using various measurement techniques. Laser devices recording flow heights and a special force plate measuring normal and shear basal forces, as well as load cells for impact pressures, were installed along the test slope. In addition, load cells were built in the support ropes and retaining cables of the barrier to provide data for detailed back-calculation of load distribution during impact. A release mechanism simulating the sudden failure of the slope was designed such that about 60 m^3 of mixed earth and gravel saturated with water can be released in an instant.

The analysis of cable forces coupled with impact pressures and velocity measurements during a testing series now allows one to develop a load model for the barrier design. The first numerical simulations of the impacted barrier lead to structural improvements of new protection measures. It appears that special adaptations to the system, such as smaller mesh sizes, a special ground-barrier interface compared to normal rock-fall barriers and channelized debris flow barriers, are necessary in order to improve the retention capacity of shallow landslide barriers.

WIT Transactions on Engineering Sciences, Vol 67, © 2010 WIT Press
www.witpress.com, ISSN 1743-3533 (on-line)
doi:10.2495/DEB100141

Detailed analysis of the friction coefficient in relationship with pore water pressure measurements gives interesting insights into the dynamic of fluid-solid mixed flows. Impact pressure dependencies on flow features are analyzed and discussed with respect to existing models and guidelines for shallow landslides.

Keywords: shallow landslide, basal forces, impact pressure, protection barrier.

1 Introduction

Landslides are gravity driven flows including rock fall, debris-flow, deep-seated landslide and shallow landslide. Shallow landslide refers to slope failure with a depth of the sliding surface up to 2 m [1]. They can mobilize up to 200 m^3 of water saturated soil material and debris. Most of the time they take place during heavy rainfall, thus their initiation is very much influenced by the structure and composition of the soil layers. Typically the presence of low permeability bedrock close to the ground surface enhances the risk of failure. The vegetation type and distribution within the soil layer will also play an important role in the stability of the slope [2].

In contrary to deep-seated landslides that are principally slow and creeping mass movements, shallow landslides release and come to a rest within tens of seconds. They are quite unpredictable and no measures can be taken during their occurrence. In spite of their limited volume compared to other phenomenon like debris-flows they can be very destructive due to their high bulk density of and to large front velocities. Habitations, roads and railway lines in the vicinity of steep terrain are primarily concerned with the shallow landslide hazard.

Up to now abundant research was done on the shallow landslide phenomena addressing various aspects of the initiation and flowing processes. The presence of pore water in the ground was studied regarding soil permeability, soil porosity and flow rates. Concepts like pore water pressure, soil suction and effective normal stress were introduced to assess the stability of slopes [3–5].

Many efforts were made to understand the rheology of landslide material and processes that condition the ability of the material to flow. The complexity of material made of particles from μm sizes like clays up to cm sizes like gravel passing over silt and sand made the task very difficult. Laboratory works including triaxial compression tests, rotating drum and small-scale chute experiments [6] were carried out in order to define viscosities and yield stresses values. However the application of results obtained in laboratory for the modelling of full-scale flows turns out to be thoughtful.

The lack of data on shallow landslides motivated the present project. The goal consisted in gathering information on full-scale shallow landslides. Flow features were to be acquired under controlled conditions in a repeatable way. The release of material down a slope makes up a good solution as long as the material composition and released volume are representative for shallow landslides.

The second goal was to investigate the impact of shallow landslides with flexible protection barriers. The measurements obtained were to allow refining and calibration of load models used in this field.

Figure 1: Test site with filled barrier, lower and upper support ropes with built in brake rings and force cells.

Figure 2: Filled release apparatus and trap door.

2 Test site

The selected testing site is a disused quarry located in the Veltheim community in the canton Aargau (Switzerland). The test slope is an 8 m wide and 41 long channel with an average inclination of 30°. The sides of the channel are about 1 m high and the bed surface is made of bedrock covered by sediments. At the top of the slope a release apparatus was built. It consists in a 1.8 m high wall whose 0.8 m lower section is a trap door that can be opened per distance. The lateral sides as well as the bottom surface above the wall are reinforced and made impermeable. The release apparatus has a capacity up to 50 m^3 material.

The landslide material is prepared by a digger out of earth material and gravel. The largest cobbles have a size up to 20 cm. Water is added until saturation and the whole is stirred up into a homogenous mixture. The material is then transported per truck and poured into the reservoir. The duration between material mixture and release lasts between 2 to 3 hours preventing material sedimentation in the release apparatus.

The flexible barrier is installed at the end of the 40 m long channel. It consists of three fields between the posts with a maximum span width of 5 m. The 3.5 m high posts are hold upslope with retaining cables. From post to post support ropes at the top and bottom hold the SPIDER wire mesh. This wire mesh consists

Figure 3: Impact pressure sensor, force plate and force cell.

of high tensile wire with mesh sizes of 130 mm that holds back most of the largest particles. Additionally a second layer of a chain-link mesh with smaller meshes prevents large draining of finer material. All support ropes are equipped with brake elements which get elongated under increasing load level.

2.1 Measurement devices

The main goals of the instrumentation and data acquisition on the Veltheim test site were on the one hand to quantify the full-scale shallow landslides in terms of velocities, flow heights, impact pressure and basal forces, on the other hand to measure the loading and deformation of the flexible barrier under impact. Several measurement devices were installed along the channel or built in the flexible barrier.

Laser distance sensors were located 14 m and 26 m downward from the release apparatus. At the second location, 2 distance sensors were placed next to each other in order to obtain 2 similar signals slightly shifted in time.

A square-shaped force plate with 0.5 m^2 surface was mounted in the channel bed surface 26 m downward from the release apparatus (location 2 of distance sensor). It measures shear and normal basal forces at the flow bottom. The force plate was enclosed in a 2 m x 1.5 m concrete foundation built flush to the channel bed surface.

Impact pressures are measured 30 m downward from the release apparatus. Two obstacles with 12 cm x 12 cm and 20 cm x 20 cm measuring surfaces were installed in the middle of the channel. The heights of the obstacles are respectively 18 cm and 26 cm.

A total of 4 force cells were built in the barrier upper and lower support ropes as well as in 2 retaining cables. They can measure forces up to 200 kN. After the release the filled barrier is measured using a tachometer. Single mesh nodes, shackles and posts are recorded with 3D coordinates.

All the data from the measurement devices were recorded with 2 kHz acquisition rate. The results in section 3 are smoothed with a moving average method over 0.1 s time intervals.

3 Results

In this section, the results from 6 experiments conducted between September 2008 and October 2009 are presented with release of 50 m^3 material. Not all

Table 1: Summary of experiments #2 to #8.

Experiment	Mean velocity (front velocity at location 1/2) (m/s)	Maximum flow height at location 2 (m)	Density (kg/m^3)	Water content (% of mass)	Main component
#2	9.5 (9/11)	0.4	1900	14	Gravel
#4	6.9 (8.6/6.1)	0.5	1850	22	Sand + fines
#5	8.7 (8.9/9.1)	0.35	1920	21	Gravel + sand
#7	9.8 (9.1/11.1)	0.3	1760	17	Gravel +Sand
#77	9.5 (10/11.1)	0.3	1760	17	Gravel +Sand
#8	7.9 (8.3/8.3)	0.35	1840	25	Sand + fines

devices were installed or worked properly at the time of the experiments so that the data available varies from one experiment to another.

The mean velocity and the amount of deposited material were not only dependant on the material composition but also on the channel bed surface. If the channel was dry and covered with sediments like in experiment #4 the flow was slower and large amount of material deposited. If the bed surface was wet or covered with little sediment the flow was faster and little material deposited.

3.1 Flow heights

The flow heights at location 1 and 2 are plotted in Figure 4. As a first remark maximum flow heights are larger for slower flow with same starting volume. The maximum flow heights decrease between location 1 and 2 when the flow front is accelerating (experiments #2, #5 and #7) and increase when the flow front decelerates (experiment #4). It shows that the flow is either spreading (maximum acceleration at the front) or compacting (minimum acceleration at the front). This interpretation of the flow height changes is consistent with the flow surface velocities and friction coefficients discussed in section 3.2 and 3.3.

In addition to the deformation of the bulk (spreading or compaction), material deposition takes place all along the channel. The deposit heights measured at location 1 and 2 vary between 5 and 25 cm. If the flow velocities over time at location 1 and 2 were known it would be possible to integrate the volume of material passing by and though quantifying the volume of the material deposited between location 1 and 2. In the present case only qualitative flow surface velocities over time are available (see section 3.2) so that it is not done. But an interesting feature of the deposition process can be observed when considering the passing time of the flow tail at location 1 and 2. The passing times coincide approximately indicating that material is coming to a rest at location 1 and probably also at location 2 Otherwise it would mean that the tail is infinitely fast between location 1 and 2. We conclude that material deposits continuously at the tail. The depositing at the tail can be recognized when looking at the basal force measurements in section 3.3.

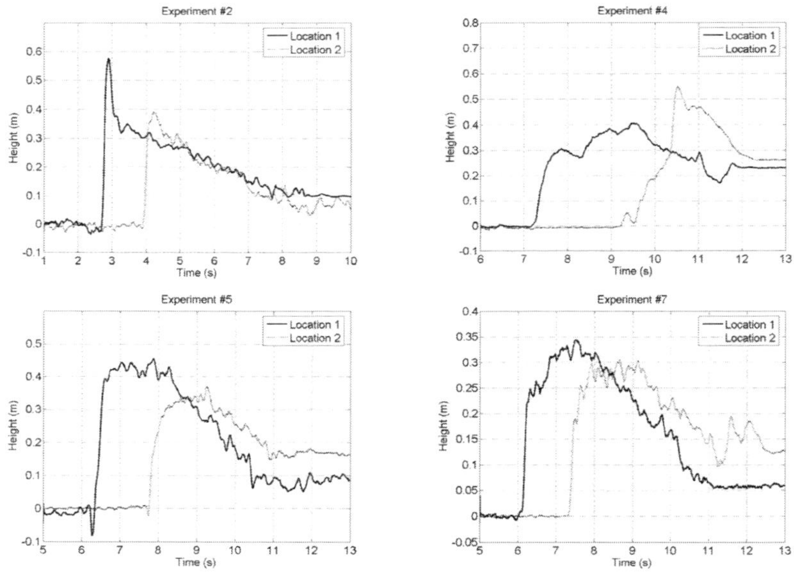

Figure 4: Flow heights versus time at locations 1 and 2 for experiments #2, #4, #5 and #7.

3.2 Flow surface velocities

The flow front velocities are easily determined from camera recordings or from the flow height measurements. Thus they provide no information about the deformation of the flowing material or about the accelerations of the rest of the flow. Flow surface velocities over time were estimated by cross-correlating 2 flow height signals recorded at location 2. The 2 distance sensors were set up next to each other with 3.5 cm distance between them. The discrete cross-correlation function used was the following:

$$h_1h_2(t,\Delta t)=\sum_{t'=t-bin\_size}^{t+bin\_size}\frac{h_1(t')-\overline{h}_1}{\sigma_{h1}}\frac{h_2(t'+\Delta t)-\overline{h}_2}{\sigma_{h2}} \quad (1)$$

where $\overline{h}_i$ is the mean value of h_i and σ_{hi} the standard deviation of h_i over the interval $[t-bin\_size, t+bin\_size]$. The time delay at time t between the 2 flow height signals is the time interval Δt that maximizes the cross-correlation function $h_1h_2(t,\Delta t)$.

The results for the flow surface velocities are plotted in Figure 5. The calculation was done for flows that were spreading. Consistently all flow surface velocities obtained all show a monotonous decrease from front to tail. No more precise conclusion can be made even if the decrease seems to be more pronounced in the tail.

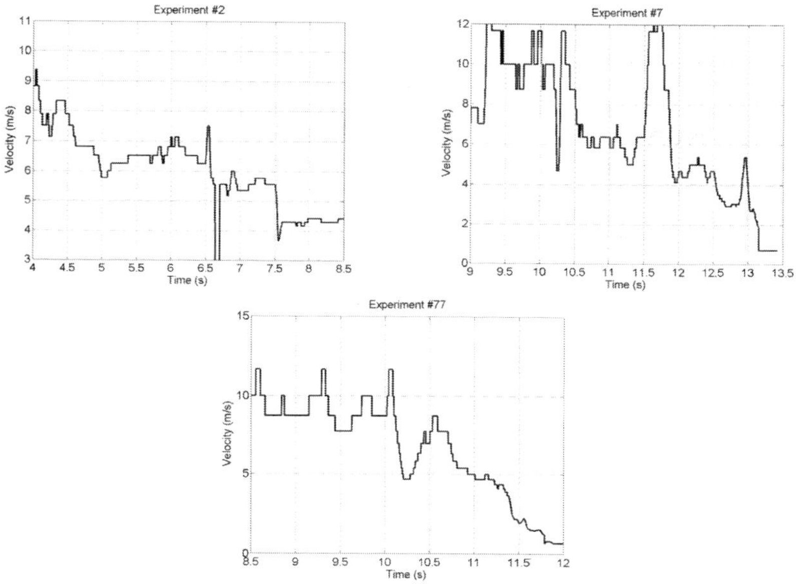

Figure 5: Flow surface velocities versus time for experiments #2, #7 and #77.

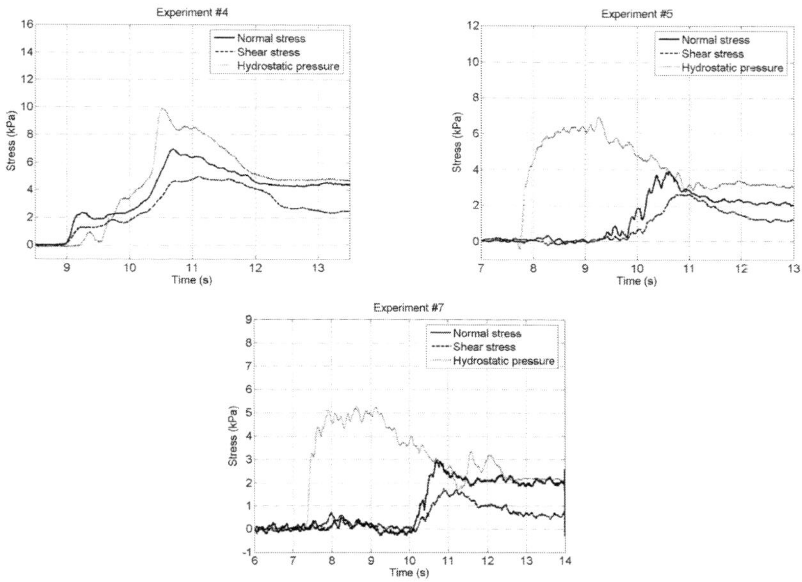

Figure 6: Basal stresses and hydrostatic pressure versus time for experiments #4, #5 and #7.

3.3 Basal forces

Normal and shear forces were measured over time with the force plate. They are represented in Figure 6 for experiments #4, #5 and #7. In the case of fast flows the problem arose that material was deflected before the force plate due to a slightly lower inclination at this location. The problem was less pronounced in the case of slower flows and inexistent for slow flow like experiment #4. The correspondence between hydrostatic pressure computed from the flow height measurements and the normal force is satisfying although the normal forces are often slightly lower. Discrepancies can be explained by the fact that laser distance sensors measure the flow height over a 0.2 cm^2 surface while the force is measured over a 0.5 m^2 surface.

Taking the ratio of the shear force to normal force gives the friction coefficient that is plotted in Figure 7. Except for experiment #4 where the friction coefficient is available over the whole flow, friction coefficient values are principally obtained for the flow bulk and tail. In the front and in the bulk the friction coefficient is higher than the tangent of the slope angle for compacting flow (experiment #4) and lower for spreading flows (experiment #5 and #7). For all experiments the highest friction coefficient values are attained in the end of the tail just before the flow height gets to a constant value (deposit height). Only then the friction coefficient value decreases to the tangent value corresponding to the equilibrium of the immobile deposit. The friction coefficient high values at the end of the tail suggest that the material is depositing over the force plate and

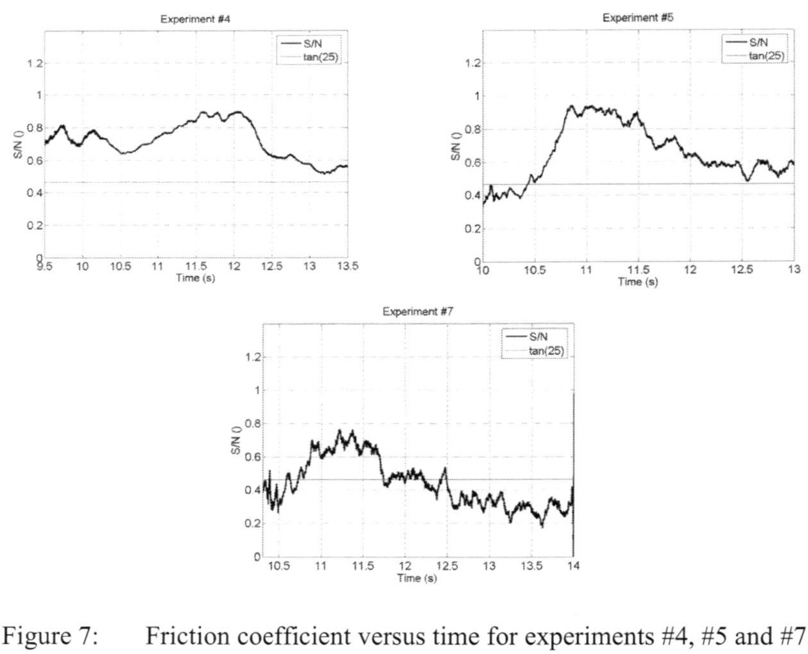

Figure 7: Friction coefficient versus time for experiments #4, #5 and #7.

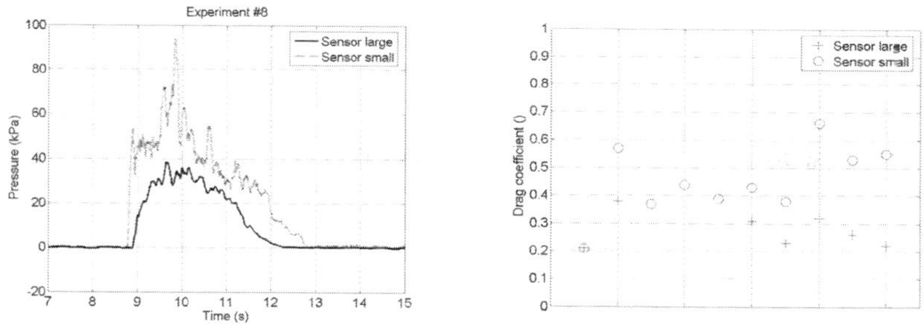

Figure 8: Impact pressures for experiment #8 and drag coefficients for experiments #4 to #8.

not just flowing past. What makes the flow stop at the end of the tail and not in the front can be due to the small flow heights at the end of the tail that would not be able to compensate some basal friction or material yield stress.

3.4 Impact pressure

Impact pressures were measured 4 m downward from the force plate. An ideal dynamic pressure measurement is not supposed to disturb the flow. In the present case the obstacle size has the same order of magnitude as the flow height. The flow was therefore deviated over and on the sides of the obstacle.

Assuming the impact pressure proportional to the material density and the impact velocity square the drag coefficient c_w is defined (see (2)). In Figure 8 the drag coefficient is calculated at the flow front for experiments #4 to #8.

$$c_w = \frac{P}{\rho v^2} \tag{2}$$

The smaller cell measures because of size effects due to particle sizes higher impact pressures leading to higher drag coefficients (see Table 2).

For further investigations of the impact pressure exerted on the barrier the values of the larger load cell are considered which seems to be more reasonable.

3.5 Interaction shallow landslide impact – flexible ring net barrier

A particular ground adaptation is necessary for the flexible shallow landslide barriers compared to the flexible debris flow barriers having a special basal opening (see Figure 9, [7]). This mesh fixed to the ground helps keeping the lower support rope to the bottom during the impact [8]. Hence also flows with small flow heights can completely be stopped and only the not innocuous liquid is able to pass through the mesh.

After 8 tests with varied mixtures, partly with several releases, we are able to present the following results for barrier design. First calculations of barrier design are given in [8]. Most important load case for barrier design is the dynamic impact of the surge and the static load case for the expected filling height.

With the measured drag coefficient for each test the intensity of the flow hitting the barrier can be estimated. For an engineering approach we assume the flow acts equally over the channel width to the bottom support rope (see Figure 10, [7]).

Assuming a constant pressure acting to the middle field and to one third of each border field (see Figure 10) the following load distribution to lower support rope can be done. The middle part of the rope is hit by the pressure q1 depending

Table 2: Drag coefficients for experiments #4 to #8.

Experiment	c_w small cell () first/second shot	c_w big cell () first/second shot
#4	0.21	0.21
#5	0.57	0.38
#6	0.37	-
#7/#77	0.44/0.39	-/-
#8/#88	0.43/0.38	0.31/0.23

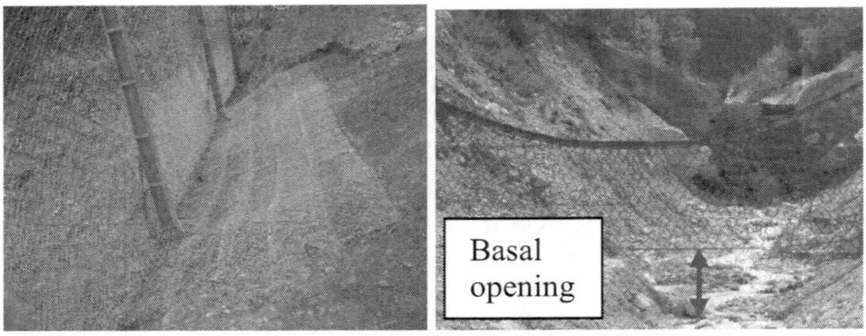

Figure 9: Particular ground adaptation with a fixed mesh for shallow landslide barriers (left) and extra projected basal openings for debris flow barriers (right).

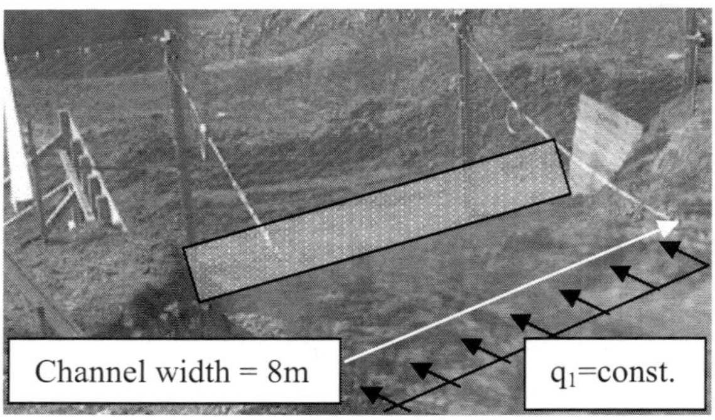

Figure 10: Impacted area on the flexible barrier and assumed constant pressure distribution in the middle field.

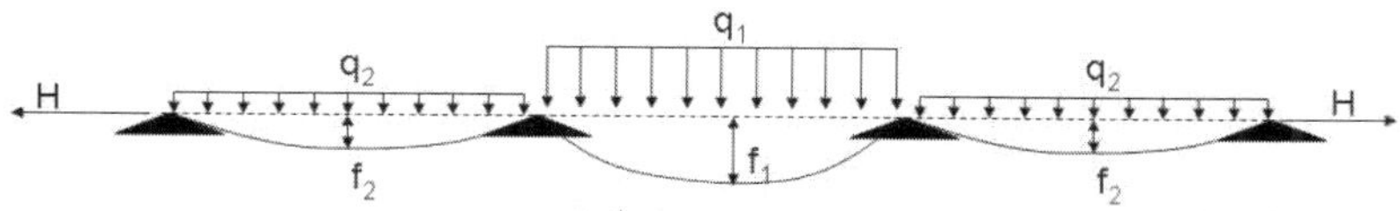

Figure 11: Simplified rope distribution to a three field supported rope.

on the flow velocity v squared, the density and the flow height given in [7]. In the border field in reality only one third of the span width is affected by the flow. This effect is considered approximately by assuming the pressure q1 over the influenced length and transferring it to the complete span width to q2 by $q_2=q_1/3$. This method allows us a simplified solution by Newton iteration of the rope equation (see (3)).

$$H^3 + H^2 \cdot EA \cdot \left[1 - \frac{1}{s_0} \cdot l\right] = \frac{EA}{2S_0}\left(\int_0^{l_1} Q_1 dx + \int_0^{l_2} Q_2 dx + \int_0^{l_3} Q_3 dx\right) \quad (3)$$

with $\int Q_i = \frac{q_i^{\,2} \cdot l_i^{\,3}}{12}$

In this assumption q_2 and q_3 are in the same order of magnitude (see Figure 11).

Last test performed without the mesh adaptation at bottom support ropes was test number three. Running back calculation of Test number three we can proof this rope equation approach without having an additional force component going into the fixed mesh to the ground. For experiment #3 we assume a drag coefficient of 0.3 which is a middle value of the measured values of experiments #4 to #8 for big cell (see Table 2). The observed flow height of experiment #3 close to the barrier was 0.15 m, the middle front velocity 11 m/s with a flow density of 2050 kg/m^3. This results into a maximum dynamic pressure of $q_1=0.15 \cdot 2050 \cdot 0.32 \cdot 11^2 = 11.9$ kN/m in the middle field. The border fields were loaded with $q_2 = 3.6$ kN/m. No brake ring deformation was observed. The rope equation leads to a rope force of 98 kN whereas the measured rope force was 110 kN. With 10% deviation the rope equation approach fits the measured rope forces. This allows us to estimate the force transmitted by the additional fix point of mesh to the ground. The calculations showed us, that the impact pressure is transmitted in the following experiments #4 to #8 half to the ground adaptation and half by the lower support ropes (see Table 3).

In four times we underestimate the measured rope force of bottom rope with maximum 8% of deviation. One experiment, #5, we overestimate the rope force with this approach. The explanation for the overestimation in experiment #5 can be a tighter installed mesh to the ground or a fault in drag coefficient or velocity values.

Of course this solution fits only for the first impact to net to lower support ropes. But the upper support ropes had always a similar load distribution mostly with a peak load of 10% lower than the bottom support ropes (see Figure 12). For the peak forces of retaining ropes the second, third release is decisive and of course the hydrostatic pressure of material behind the barrier [8].

Table 3: Comparison of the solution of the rope equation with measured rope force.

Experiment	Impact to middle field (kN/m)	Solution of rope equation (kN)	Measured rope force (kN)	Deviation (%)
#4	1.8	11	12	-8.0
#5	16.6	76	79	-3.7
#6	36.5	116	98	+18.0
#7	17.0	83	85	-2.3
#8	10.6	53	55	-3.6

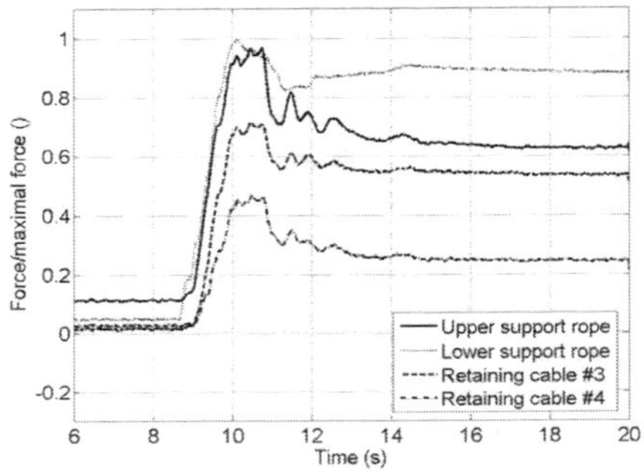

Figure 12: Force to maximal force ratio versus time in the upper and lower support ropes and in retaining cables #3 and #4.

4 Conclusions

The results of full-scale shallow landslide experiments are presented. 50 m^3 of landslide were released on a 40 m long and 8 m wide channel with 30° inclination. The material was made of gravel, sand and clay saturated with water. Flow heights, basal stresses, front velocities and surface velocities were measured. At the end of the test slope a flexible barrier with high tensile steel net installed. Forces in the support ropes and retaining cables were recorded during impact.

The flow height measurements allowed the distinction between spreading and compacting flows. The time interval between the passing time of the tail ends suggested that material is deposited continuously at the tail end. Cross-correlation of flow height signals showed monotonous decrease in the flow surface velocity for spreading flows. The analysis of basal stresses revealed a systematic maximum of the friction coefficient in the tail end consistently with the interpretation of the flow height measurement. The friction coefficient values in the front and in the bulk depend to a large extent on the bed surface friction properties.

A method was developed to estimate the maximum dynamic load in the barrier during impact of the flow. By assuming constant drag coefficient over the channel width the forces in the lower support rope were calculated and compared with the measured forces solving the rope equation. The effect of the ground adaptation was also estimated.

References

[1] Kantonale Gebäudeversicherungen (2005), Objektschutz Rutschungen.

[2] Schwarz M. (2008), Characterization of the vegetation cover at the test site of Rüdlingen, *Internal CCES-TRAMM report*.

[3] Terzaghi K. (1925), *Erdbaumechanik*, Franz Deutike, Vienna.

[4] Iverson R.M. (1997), The physics of debris-flows, *Reviews of Geophysics*, 35, 245-296.

[5] Springman S. et al. (2009), Landslide triggering experiment in a steep forested slope in Switzerland, *Proceedings of the 17th International Conference on Soil Mechanics and Geotechnical Engineering*, doi:10.3233/978-1-60750-031-5-1698.

[6] Iverson R. M. et al. (2009), The Perfect Debris Flow? Aggregated Results from 28 Large-scale Experiments, submitted.

[7] Wendeler C. et al. (2010), Structural design of flexible steel barriers for torrent debris flow mitigation, in preparation.

[8] Bugnion L. et al. (2008), Versuche zum Rückhalt von oberflàchennahen Rutschungen mit flexible Schutznetzen. *Proceedings of the Christian Veder Kolloquium*

Landslide in a catchment area of a torrent and the consequences for the technical mitigation concept

F. J. Riedl
Austrian Federal Service for Torrent, Erosion and Avalanche Control, District Office "Middle Inn Valley", Austria

Abstract

The focus of the following lies on the identification of the practical solutions by taking into account the theoretical background. A translational landslide occurred on October 2008 within an area of about 2.5 ha on the topographical left side of the torrent "Wattenbach" in Tyrol/Austria.

In the summer of 1965 a large debris flow event of the torrent "Wattenbach" happened and the underlying city of Wattens was completely destroyed. After this event several technical protection measurements were implemented to guarantee a certain factor of safety for the city of Wattens.

After the landslide event of 2008, the most important question concerned the interaction of the torrent "Wattenbach", the landslide and which reaction could be expected by a flooding event in the future.

To obtain certain quantitative and qualitative data, several analyses (modelling of the landslide by regarding different scenarios, laser scans, field works) were implemented and some of them are still going on.
Keywords: landslide, debris flow, torrent, Austrian Federal Service for Torrent, Erosion and Avalanche Control, technical protection measurements, natural hazards.

1 Introduction

1.1 Austrian Federal Service for Torrent, Erosion and Avalanche Control, District Office "Middle Inn Valley"

The main tasks of the Austrian Federal Service for Torrent, Erosion and Avalanche Control are divided into hazard zone planning (risk prevention),

WIT Transactions on Engineering Sciences, Vol 67, © 2010 WIT Press
www.witpress.com, ISSN 1743-3533 (on-line)
doi:10.2495/DEB100151

planning of technical mitigation measures against natural hazards, building constructions works and expertises for the public authorities.

The District Office "Middle Inn Valley", situated in Innsbruck, is responsible for the torrent, avalanche, rockfall and erosion protection in the two districts Innsbruck-Town and Innsbruck-Country. The total area covers 210.000 ha whereby only 15% of them can be used for permanent settlement.

There are also 256 torrent catchment areas and 264 harmful avalanche tracks which endanger the permanent settlement. In the year 2009 the monetary investments were about 5.0 million € in the preventive technical, forestry and soil-bioengineering measurements.

2 Landslide event on October 2008, city of Wattens, district of Innsbruck-Country

2.1 Kinematic analysis of the landslide in the catchment area of the torrent "Wattenbach"

The landslide in the catchment area of the torrent "Wattenbach" occurred on October 2008 on the topographical left side of the torrent. For analysing purposes, a field work was conducted to obtain data about the geology, the geomorphology, the level of the mountain water, the surface runoff and the initial structural situation (brittle and ductile deformation, foliation, etc).

2.1.1 Geology

The geology is defined by a fine foliated phyllite, the so-called "Innsbrucker Quarzphyllite". In general, the northeast-exposed hillsides are ancient deep seated gravitational landslides which are nowadays in a nearly firm stage [1]. From the structural and tectonical point of view the phyllite was deformed by several different phases from D1-D4 [2]. Actually, a steady influence of the

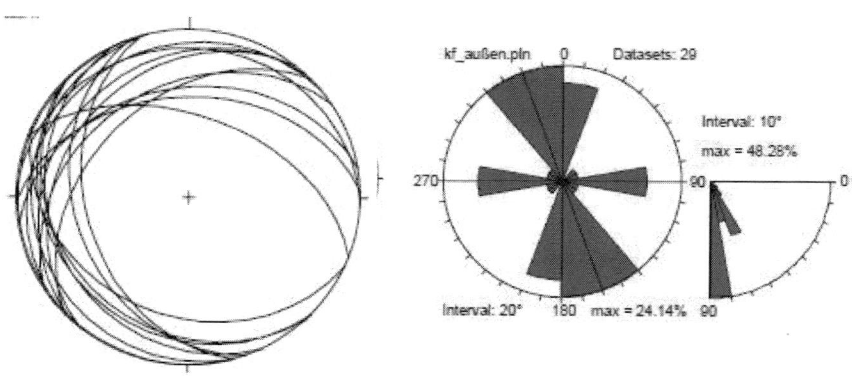

Figure 1: Orientation of the foliation and the brittle cracks within the landslide [3].

Figure 2: Geomorphological mapping and a profile through the active translational landslide [3].

active brittle sinistral Inntal fault can be examined which causes a negative structural influence on the stability of the preliminary deformed phyllite.

In this specific case the dip of the foliation diversifies, which is caused by a singular rotation, from south to west. Furthermore a nearly constant dip of the foliation to the southwest can be recognized. The brittle deformations within the landslide are characterized by steep northwest-southeast orientated cracks [3].

2.1.2 Geomorphology

From the kinematical point of view the landslide is a translational slide [4], partly within the solid rock, with a depth of the slip plane of about 10-20 m. As mentioned before the active landslide is part of an enlarged ancient deep seated gravitational landslide system. The torrent "Wattenbach" has eroded the convex front of the moving part during the last centuries and since the catastrophically debris flow event of 1965, the toe of the slope has become precipitous. The main scarp dips 50-70° to the northeast and on the topographical right side of the boundary zones the dip of the scarp rotated to the north.

2.1.3 Hydrogeological and hydrological runoff

There are two diffuse hydrogeological mountain water zones, one between 850-860 m a.s.l. and the other one between 810-820 m a.s.l. The hydrological surface

Figure 3: Debris flow event in August 1965 and the damages in the middle part of the torrent and the height of the water level by the bridge in the city of Wattens (above).

Table 1: Results of the slope stability analysis.

Szenarien	Kombination 1	Kombination 2	Kombination 3
a1	1,0	1,0	1,0
b1	1,0	1,0	0,90
b2	0,83	0,81	0,83

Tabelle 1: Ergebnisse für die Berechnungen mit Geländeprofil 1

Szenarien	Kombination 1	Kombination 2	Kombination 3
a2	1,21	1,14	1,10
c1	1,10	1,10	1,08
c2	0,98	0,98	1,05
d1	1,21	1,21	1,10
d2	1,21	1,21	1,10

Tabelle 2: Ergebnisse für die Berechnungen mit Geländeprofil 2

szenarios

a1	development of the landslide topographic profile 1
a2	development of the landslide topographic profile 2
b1	influence of the mountain water zone 1 by profile 1
b2	influence of the mountain water zone 2 by profile 1
c1	decrease of the riverbed 5,0m profile 2
c2	decrease of the riverbed 10,0m profile 2
d1	increase of the riverbed 5,0m profile 2
d2	increase of the riverbed 10,0m profile 2

runoff is marked by several small gullies with some initial erosion. Local technical mitigation measures have been implemented by the Austrian Federal Service for Torrent, Erosion and Avalanche Control Service, District Office Middle Inn Valley on a tributary to the "Wattenbach".

The catchment area of the main torrent "Wattenbach" is 74 km² with a peak runoff of about 90m³/s with a regarded repeat period of time of 150 years. On the base of the hazard zone planning, the expected bed load is about 160.000m³. The debris flow event of 1965, where large areas of the city Wattens were destroyed, the measured bed load was about 85.000 m³ [5].

The main drinking water spring of the city Wattens is above the active landslide on 920 m a.s.l. and the remaining water was currently flowing uncontrolled into the active moving zones.

2.2 Geotechnical investigations and modelling

The main tasks of the geotechnical investigations were to examine the development of the landslide, the role and importance of the mountain water level and the development of the landslide by decreasing and increasing the riverbed of the torrent "Wattenbach". The numeric modelling of the landslide, by regarding the topics, mentioned above, was done by an extern consulting engineering company with the finite element software PHASE2, Rocscience Inc. The results of this modelling should be the base for the further technical mitigation concept.

By analyzing the different scenarios (two topographical profiles and the influence of the two zones of the mountain water levels) the results of the slope stability analysis are as follows.

The main conclusion of these geotechnical investigations and the results of the varying scenarios were the fact that a decrease of the riverbed of the torrent "Wattenbach" with more than 10,0 m would lead to a slope failure (c2). This

Abbildung 42: Grenzzustand des Hanges (Deformationen farblich dargestellt)

Figure 4: Modelling of the landslide and the deformation by regarding the different scenarios.

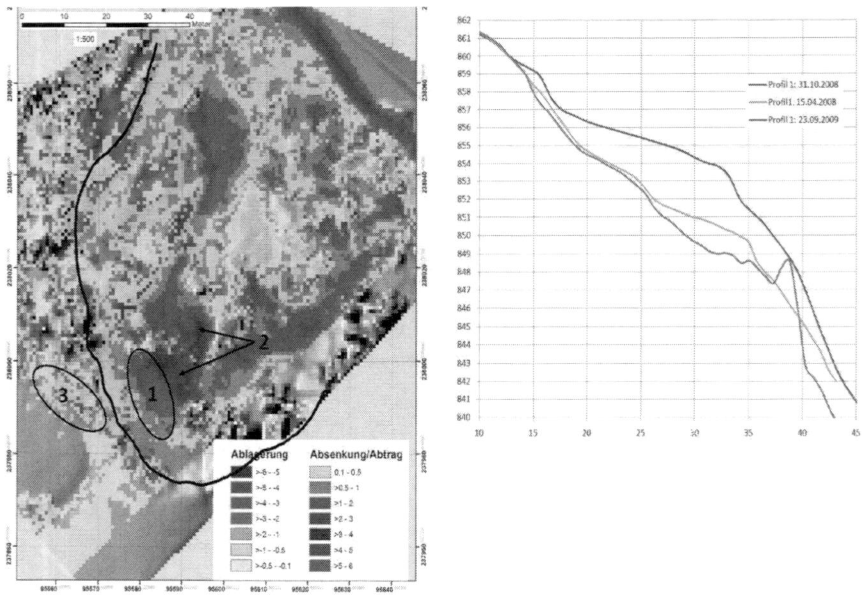

Figure 5: Difference in height inside the landslide and the distinctive depression in the upper part [7].

induces to a technical mitigation concept whereby the riverbed has to be fixed against depth erosion. An increase of the riverbed results in a minor increase of the factor of safety (d1-d2). Nevertheless the mountain water level is fundamental and has to be considered for the further mitigation planning [6].

2.3 Laser scanning of the landslide

In the alpine regions the original use of the laser scanning is based on the examination of the snow cover. The scanning of a surface of a landslide and the obtained experiences for the future was one of the defined tasks. A cooperation partner, the Federal Research and Training Centre for Forests, Natural Hazards and Landscape, has fulfilled on base of a cooperation contract the laser scanning of this landslide. The scanning was done with the Scanner LPM98-2K by the company Riegl with the highest cancelation and a projected spot spacing of 0,75m [7].

The focus of the laser scanning investigation was to obtain compressions and depressions within the active zones and further detailed information about the kinematical movements. It is also important to point out the fact that the technical mitigation achievements should be quantified after finishing the construction works for a defined period of 3-4 years. The reference measurement has already been created after the event 2008 and the main confiding at this time were the enormous depressions on the upper part of the landslide up to 6,0m.

3 Technical mitigation concept

3.1 General conspectus

On the base of the several investigation results, the technical mitigation concept was according to them.

The main conclusion of the slope simulation was the fact that the riverbed should be fixed and consolidated to obtain slope stability. Due to the induced depth erosion, caused by the debris flow event in den "Wattenbach" in the summer of 1965, a collapse of the slope stability could occur and additionally bed load material could be transported up to the city of Wattens.

3.2 Consolidation of the riverbed of the torrent "Wattenbach"

For the stabilisation and consolidation process of the riverbed of the torrent "Wattenbach", at least 13 check dams with a certain debris flow section will be constructed to avoid depth erosion during a debris flow event. The dimension of the check dams will be designed on the peak runoff of $90m^3/s$. The needed height of the check dams is about 3-4m, the concrete cubature is about $300m^3$ and the steel demand is about 8,5 tonnes per check dam. The toe of the landslide should be firm up by stabilizing the riverbed through these check dams.

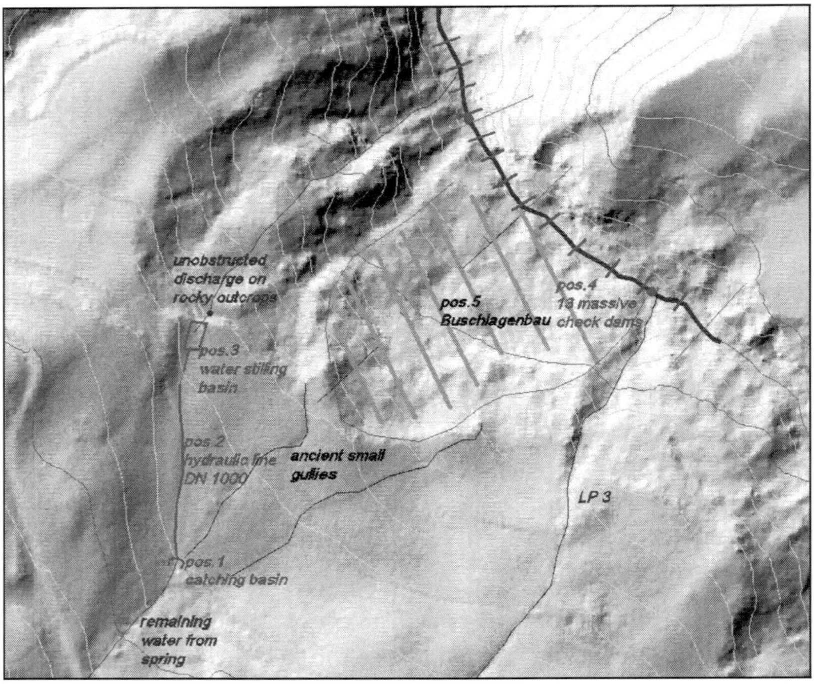

Figure 6: Technical mitigation concept at the basis of the investigations.

3.3 Hydrological surface runoff of the small gullies

The small gullies beyond the landslide, which infiltrated into the instable slope, will be displaced with an earth-covered hydraulic line DN1000 to the rocky outcrops on the topographical left side of the landslide.

To decrease the kinematic force of the water, a stilling basin will be constructed on base of the hydraulic dimension. Due to the rocky outcrops the surface runoff of the small gullies can then be discharged unobstructed on them.

3.4 Soil-bioengineering measurements

The diffuse characteristic of the two mountain water zones within the landslide will be conceived by soil-bioengineering measurements, the so-called "Buschlagenbau" [8]. Differential salix are used for the "Buschlagenbau" and they will be orientated to the topographical left rocky outcrops, to achieve a two-dimensional drainage effect. Another aspect of the "Buschlagenbau" is a small area stabilisation of the upper soil-complex. These soil-bioengineering measurements are the highly recommendable measurements due to uncontrolled diffuse water flow. The main mechanical movements of an unstable slope are in general caused by such diffuse water flows.

3.5 Perspectives

To obtain the expected achievements of this technical mitigation concept it is mandatory to make further investigations in analyzing the landslide with the laser scanning after finishing the construction works. The quantitative and qualitative movements of the investigation points will give an explanation of the landslide and a confirmation if the technical measurements worked as successfully as expected.

References

[1] Hermann, S., Tiefreichende Großhangbewegungen im Kristallin der Niederen Tauern, Ostalpen. –Verbreitung, Typen und ihr Einfluss auf die Morphogenese alpiner Täler. *Geoforum Umhausen (GFU)*, **1**; Innsbruck, 1999

[2] Rockenschaub, M., Kolenprat, B., Frank, W., The tectonometamorphic evolution of Austroalpine units in the Brenner area (Tirol, Austria) – new geochronological implications. *Tübinger Geowissenschaftlichen Arbeiten, Series A*, Vol. **52**, pp. 118–119, 1999.

[3] i.n.n., Rutschung Eggerbachl - Bewertung des Ist-Zustandes und Abschätzung der Auswirkungen auf den Hochwasserabfluss im Wattenbach. *Intern Report*, Innsbruck, 2009.

[4] Varnes, D. J., *Slope Movement Types and Processes – in: Special Report 176: Landslides: Analysis and Control*, (R. L. Schuster and R. J. Krizek, eds.), TRB, National Research Council, Washington D.C, 1978.

[5] Forsttechnischer Dienst für Wildbach- und Lawinenverbauung, Wattenbach. *Kollaudierungsoperat 1993 für die Baujahre 1965-1991*, Technischer Bericht, pp. 21, Innsbruck, 1993

[6] GEC ZT GmbH, Rutschung Wattenbach/Eggerbach. *Intern Report*, Innsbruck, 2009.

[7] Federal Research and Training Centre for Forests, Natural Hazards and Landscape Department Natural Hazards and Alpine Timberline Unit Water Balance in Alpine Catchments. Kurzinformation zur Massenbewegung Wattental, Laserscanning, *Intern Report*, Innsbruck, 2009.

[8] Schiechtl, M., Stern, R., *Handbuch für naturnahen Erdbau*, Österreichischer Agrarverlag: Wien, pp. 92-96, 1992.

Regional methods for shallow landslide hazard evaluation: a comparison between Italy and Central America

D. Brambilla, L. Longoni & M. Papini
Politecnico di Milano, Department of Environmental, Hydraulic, Infrastructures and Surveying Engineering, Italy

Abstract

Landslides are a serious issue in the Latin American and Caribbean region, causing disasters and casualties every year. Since these regions suffer from endemic poverty and few resources can be allocated for civil protection, the need to maximize results is strong. In order to diminish vulnerability and increase effectiveness of any intervention, a wide knowledge of territory critical situations needs to be achieved. This paper suggests two useful tools to build hazard maps of a chosen territory; these maps can be useful as a Decision Support System (DSS) to help the management of economical resources and to identify the situations that have the greatest need. The two methods proposed are Stability Index MAPping (SINMAP) and the Mora and Vahrson method. Both methods start from the digital terrain model and other various parameters, linked mainly to slope lithology and rainfalls, and obtain a map of the relative hazard from shallow landslides. The methods have been applied to two study cases: one in Guatemala, a perfect example of what has been mentioned about the difficult Central American situation, and one in the Varese district in Italy, a well known site which provides plenty of data and information that was useful to the authors for rigorous testing of methods. Finally the results have been compared, highlighting the strong influence of input data quality on results and the interesting potential of the tested methods; these proved to be successful when input data have enough resolution and to be useful as a DSS in order to identify critical areas and intensify efforts in these areas.
Keywords: shallow landslides, Central America, Stability Index MAPping (SINMAP), Mora and Vahrson, GIS, hazard map.

WIT Transactions on Engineering Sciences, Vol 67, © 2010 WIT Press
www.witpress.com, ISSN 1743-3533 (on-line)
doi:10.2495/DEB100161

1 Introduction

Latin American and the Caribbean regions are heavily affected by natural disasters. According to a recent hypothesis, the disappearance of the Maya civilization is now believed by historians to be linked to an unusually long and severe drought. Because of its geomorphologic situation and its geographical position, the region is prone to natural events of severe intensity, although experts link the large human and economic cost associated to these events to an extreme vulnerability.

Looking back only to the last 30 years, we can count an average of 32,4 disasters per year in this region, which led to a total of 226.000 fatalities, that is to say about 7.500 per year. Analysis shows that the frequency of disasters seems to rise during the 20th century, although it is possible that this is not connected with a real increase in natural disasters, but rather with the lack of data available for the first half of the century [1].

Obviously all these disasters have meant enormous economic costs for the countries hit, not only linked to the disaster itself but also because of the weak response mechanism. The economic effects involve not only restoration, but also the disruption of economic activities with an immediate impact on the Gross Domestic Product (GDP) growth. Given that this area is exposed to natural risks and the economic resources that can be allocated for civil protection and risk forecasting are limited, the aim of this work is to investigate if any methods exist that could be helpful in gathering better territory knowledge. It will be very important to get an overall view of the territory and its critical points; moreover, these methods should be integrated into a Decision Support System (DSS) to help manage economic resources and efforts in order to maximize the effectiveness of any actions.

Testing the possibility of applying regional methods to forecast landslide susceptibility in large areas, with little effort and starting from a limited amount of data, is a basic step to encourage their diffusion and use. In order to prove that these methods can be useful for DSSs, two opposite case studies were chosen: one in Guatemala, a perfect example of what has been mentioned about the difficult situation in Central America, with little input data and extensive landslide hazards. The other one is located in the Varese district in Italy, a well known site with plenty of data and information that was useful to the authors for rigorous testing of methods; moreover, it is a good chance to apply the Mora and Vahrson method in a European context.

2 Basin scale methods

When dealing with landslide risk in such extensive areas, two main problems arise immediately: the lack of data and the small amount of time and money available to accomplish the work. So the authors focused on basin scale methods for landslide risk forecasting. Generally speaking, these methods start from geographic information in order to generate a thematic map that can give a rough evaluation of the risk. Whatever method is chosen, the results will be strongly influenced by the quality of the base data. Basically it is impossible to obtain a good result starting with rough data and this issue will be explained in the work. So, the regional method has the power to make a quite good forecasting of events, but it needs careful application and calibration work, since every different kind of terrain will react differently to the same conditions. Although these applications will require validation and a check of rightness, they have a really useful feature: they are totally automatic and the whole work can be carried out by a trained operator and a personal computer with GIS software installed on it. The authors do not suggest totally neglecting the in situ analysis and local studies, but the first step of the work does not need this information, which requires a long time and high costs to be collected. Moreover, the accuracy of the results is really effective and with a good quality to cost ratio, assuming that the model result is a thematic map and is only a good indication of risk and not an absolute evaluation.

2.1 Types of landslide

Since analyzing every single kind of landslide that could happen in a large territory is quite impossible, or at least very time consuming and expensive, the authors decided to focus only on shallow landslides, these events being the most common in the study areas. Such events involve only limited zones of terrain, with a moderate thickness, not more than 5 meters, and the moving material is made mainly of surface debris. These landslides are frequent and very dangerous, both because of the starting points, which are not easily determinable, and because of their evolving speed. When referring to shallow landslides usually geologists refer to soil slips and dumps and to debris flow.

The authors chose two different regional methods that are commonly used for regional scale shallow landslide forecasting: Stability Index MAPping (SINMAP) and the Mora and Vahrson method.

SINMAP was developed in 1998 by a team at Utah State University. It works as an extension of ArcGis software, a software by ESRI inc., and it give as a result a map of areas of potential instability and landslide risk. The model uses as input files a DEM of the studied region and a shapefile containing known points of landslide initiation. SINMAP requires the calibration of some parameters based on the information collected in situ. The model is based on an indefinite slope equation that evaluates the safety factor of an area. The equation takes into account a variety of parameters, ranging from the slope and wetness of the terrain to the soil and root cohesion [2].

The Mora and Vahrson method was developed by Mora and Vahrson, and published in TC4, *Manual for zonation on seismic geotechnical hazard.* This method considers the role of three main factors, morphology, lithology and soil humidity, on landslide susceptibly and two triggering factors: earthquakes and rainfalls. These factors are multiplied together to get a final number indicating the landslide risk level. The Mora and Vahrson method is not widely used in Europe, since it was developed and tested for the Central American situation [3].

3 Guatemala study case

The first case study chosen is in Guatemala. This country is extremely prone to shallow landslides; moreover, the resources allocated to civil protection are limited, adding a noticeable vulnerability to an already critical situation. Guatemala is located in Central America, just next to the southern border of Mexico, and it is threatened by two types of natural risks: earthquakes and landslides. The country is located at the intersection of three active tectonic plates, the Cocos, the Caribbean and the North American, consequently in the past centuries major earthquakes have destroyed communities and livelihoods. Hurricanes are the other main cause of disasters in Guatemala, provoking great impacts on the agricultural sector and road infrastructures; both earthquakes and hurricanes are triggering causes of debris flows, lahars and landslides. Such events have been reported since the Spanish arrived in Guatemala: the first written documents about big landslides date back to 1541, when on the 11th September massive rainfalls associated with a tropical hurricane triggered a large landslide that buried the city of Santiago de los Caballeros, which at the time was the seat of the Spanish Government.

Nowadays the situation is still critical, as the population of Guatemala has grown through the centuries the country has undergone heavy deforestation. Because of this landslides are very common, in particular during the rainy season stretching between March and October. Guatemala does not have the large economic resources necessary to sustain landslide research and, in addition, the input data were quite rough, with a low resolution. To evaluate the Guatemala case study others authors' works were studied and their results are now reported. The first work applied SINMAP, the second one Mora and Vahrson. A technical report performed by Rachel Chisolm, from the University of Texas in Austin, applied SINMAP to Guatemala referring results to Hurricane Stan in October 2005 [4].

The DTM used had a 30 second arc resolution and monthly average rainfalls collected by Global Precipitation Climatology Center (GPCC) were the only rainfall information used. Moreover, the author also used a map developed by the Instituto Nacional de Sismologia, Vulcanologia, Meterologia e Hidrologia (INSEVUMEH) that shows the accumulated rainfalls for the 10 day period after the start of Stan.

Unfortunately these data were too rough for a good application of the model and the authors obtained a result, showed in Figure 1, which was inadequate. This map shows the contributing area of each cell, and all the cells have a contributing area of less than ten cells. This is illogical in a country of varied topography, such as Guatemala, and is due to the coarseness of the DEM. The 30 second arc resolution is too rough to see the largest part of the topography variations. This has the effect of smoothing out the terrain, so an accurate analysis could not be made. So, it is concluded that a relationship between altitude, rainfall and landslides exists, as showed by Figure 2.

Better results were expected with the application of the Mora and Vahrson method. The work was published by Geopetrol SA in "Estudio hidro-geologico

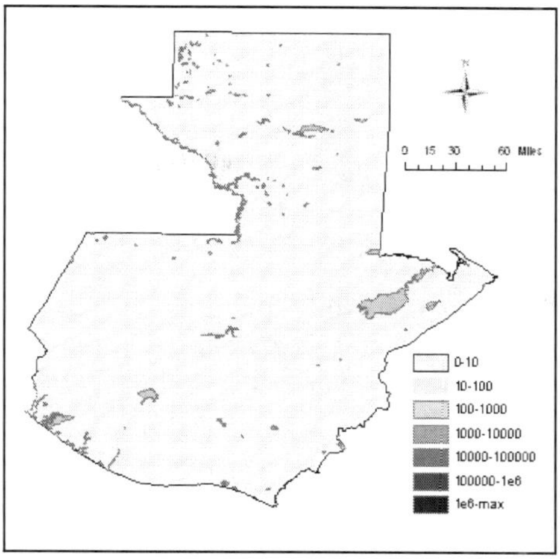

Figure 1: SINMAP contributing area and landslide locations.

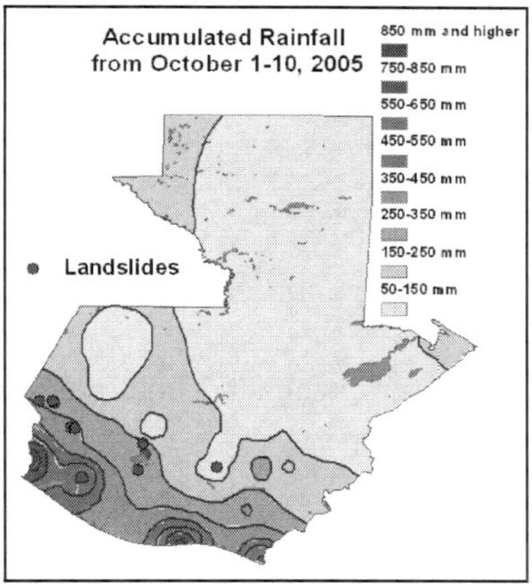

Figure 2: Landslide locations and accumulated rainfall.

para lòa implementacion de un sistema de monitoreo y alerta ante deslizamentos en asentaminetos urbanos del departemento de Guatemala, Centro America". Notice that the investigated area in this study was not the entire country, but only the central region called Zona Metropolitana de Guatemala. These results are more encouraging, mostly because the authors could rely on a slightly better topographic input [5].

Actually the DEM was built up starting from level lines of the national map. Then the method necessitated some lithological parameters as input values, which could be easily extracted from the geological map with little or no influence of scale. It is important to say that in this example earthquakes were neglected as triggering factors, because the authors wanted to focus only on rainfall triggered landslides. The map in Figure 3 represents the area of homogenous hazard divided into Low, Moderate, High, Really High. The map in Figure 4 represents the spatial distribution of landslides events in the studied region between 2005 and 2007. It is impossible to make a comparison between the results of the two methods, because of the lack of data from SINMAP. Relying only on Mora and Vahrson, the first thing that requires attention is how the landslide hazard grows moving from north east to south west. This is due to two factors: rainfall intensity, which grows exactly that way, and terrain elevation.

However, the hazard map has only a slightly correspondence with the real distribution of landslides: there is at least one main reason for that. The input DEM was not a high quality one so, starting with the evidence that the greatest

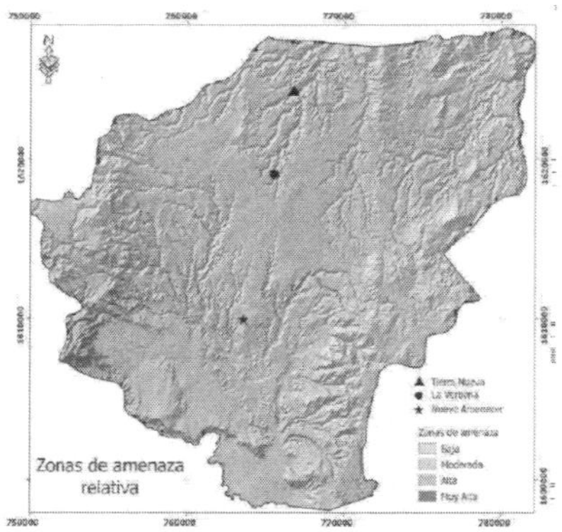

Figure 3: Hazard map landslide location in Zona.

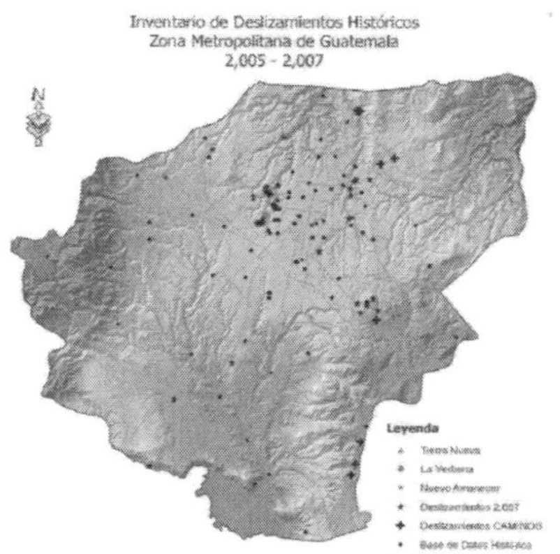

Figure 4: Metropolitana de Guatemala.

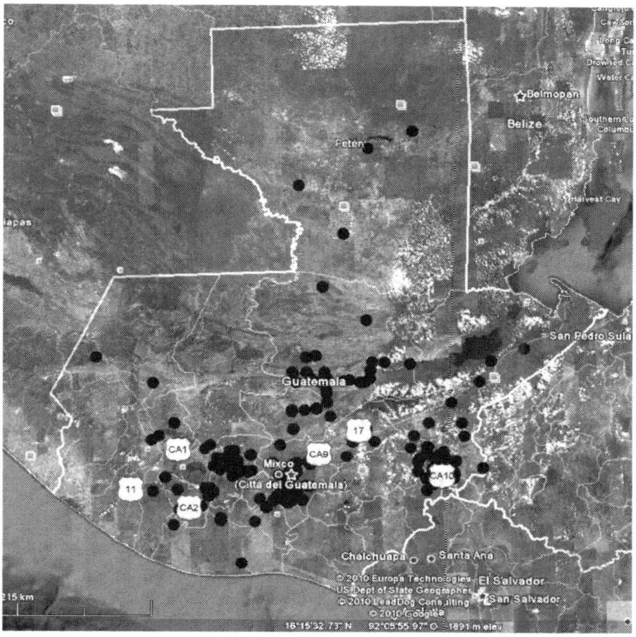

Figure 5: Guatemala landslide locations, black dots.

part of landslides happened in really narrow valleys, the terrain model cannot see them and cannot consider their slope in the calculations. The direct consequence is the impossibility to forecast these events.

It is important to have a database collecting every event that has happened in recent years, because this the paper presents efforts to compile an archive of the many events that have happened in Guatemala in the last ten years; this database is mainly based on many events collected in INSEVUMEH papers [6], which are represented in Figure 5. It is possible to notice that landslides are grouped in the southern part of the country, as was expected. So comparing the literature results with this summarizing map gives an idea of the method's ability to forecast landslides; notice how, while starting form similar data, SINMAP is not able to give a result in Guatemala, while Mora and Vahrson can also make some hazard mapping when input data are rough and with a low resolution. The inventory confirms what has been found by Chilsom, that there is a strong correlation between slope rainfalls and landslides.

4 The Italian study case

This paper applies the methods already used in Guatemala to a well known territory in Italy in order to evaluate their reliability. This second case study features a lot of data from previous studies. Moreover, this area is close to the authors' location and they could get directly on the terrain to better understand the events in progress. The authors could also compare their results with the Inventario Fenomeni Franosi Italiani (IFFI) database. The IFFI was a big project that had the ambitious objective of summarizing all the Italian landslide events in a georeferenced archive. The zone investigated is the Comunità Montata della Valganna and Valmarchirolo in the Varese district, Lombardia, Northern Italy. The surface is about 55 km^2, and it is mainly a mountain area. The terrain altitude varies from 200 m above sea level to 1100 m. This area is commonly involved in shallow landslides events. The authors applied the Mora and Vahrson method, starting from high quality input data. A 20x20 meters DEM was available and also small-scale geological maps. During data elaborations a problem arose: the Mora and Vahrson parameters are calibrated on the Central American climate, so the evapotranspiration value was not suitable for the Italian climate. The authors decided to adjust the scale of value in order to center it on real observed results. Actually the use of original scale of values would have labeled all zones as maximum value, removing sense from the parameter itself. The result of the elaboration is reported in Figure 6; the shades represent the hazard level: Low, Medium, High and Really High. The same territory has been investigated with SINMAP. In this case the results are also good; the input data were sufficiently precise and the model could work with success. SINMAP produces a thematic map where different colors correspond to different hazard levels, as shown in Figure 7. Making a comparison between the two methods is possible in this case; first of all both the models give good results, the hazardous zones are the same and so are the safe zones. Mora and Vahrson seems to suggest that only a small area is in low hazard zone, while SINMAP labels

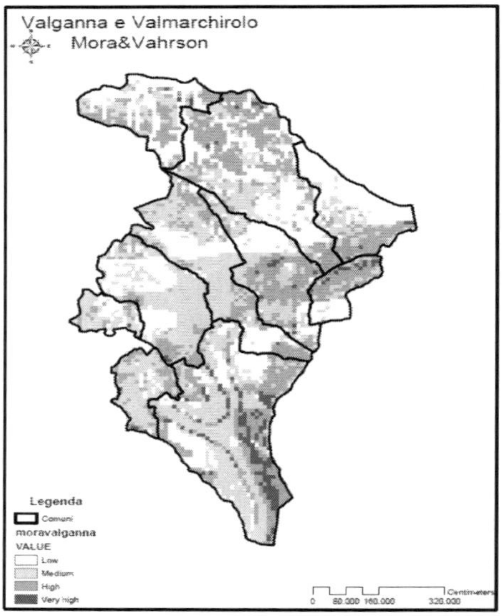

Figure 6: Valganna and Valmarchirolo hazard according to the Mora and Vahrson model.

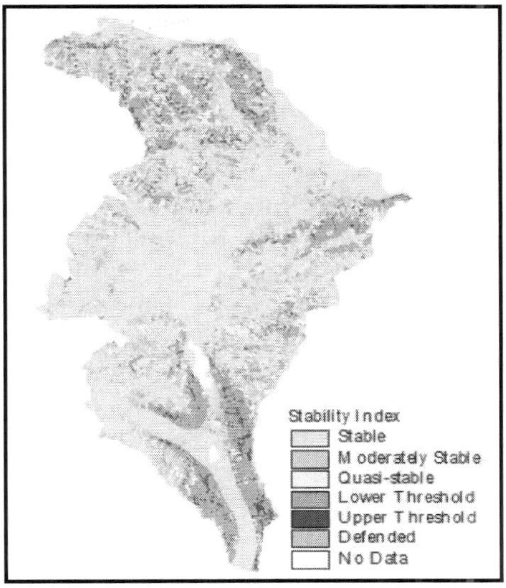

Figure 7: Valganna and Valmarchirolo hazard map according to SINMAP.

Figure 8: IFFI database of slides in Valganna and Valmarchirolo.

the biggest part of Valganna and Valmarchirolo as stable. This is probably due to a different mapping of results: SINMAP has six levels, while Mora and Vahrson has only four; moreover, the second method is very sensitive to slope, and in all the areas investigated only a marginal fraction is plain. The map in Figure 8 summarizes all the shallow landslides known in the studied zone taken from the IFFI database; it is possible to notice how all the events fall in the most dangerous areas. These areas are logically concentrated in hilly and mountainous zones, since slope is one of, if not the most, influencing factor when mapping landslide hazards. Looking at the SINMAP results it is possible to find a good correlation between calculated risk and real event locations. The model is able to predict that 70% of the reported landslides lay in hazardous zones. Finally it is clear that both the models give good results, but probably SINMAP is slightly better because it is able to indicate the instable zones, leaving a fair percentage of no risk or safe territory, while in Mora and Vahrson only a small part is marked as Low Risk. Although showed to be easily adaptable to a terrain (the Alps region), the results from this last model were completely different from those used for calibration.

5 Study case comparison

Finally, it is possible to make a comparison between the two examples chosen: Guatemala and Varese. The first thing that can be noticed is that the results are strongly influenced by input data. It is really important to have good quality data to apply the model and gather results. In the Guatemala case study the maps are poor and imprecise and despite the efforts of the authors, possible landslides cannot be correctly identified within the hazard zones. This prevents the authors from having the chance to test the methods and evaluate their reliability. Luckily, in the Italian case study, the results demonstrate that both SINMAP and Mora and Vahrson can provide useful indications about slide hazards. It is important to point out that the Italian case study was much smaller than the Guatemalan one, but this has no effect on the Mora and Vahrson model, since the same data used for Valganna and Valmarchirolo are available for the whole of Italy. Unluckily, but logically, SINMAP requires really precise data in order to work; these data need to be collected directly on the terrain. For example, the model needs information about climate and root cohesion, not in terms of a numeric evaluation but as a probability distribution. Moreover, it is strongly conditioned by the landslide triggering point database. These data helps the model to be reliable, but are not easily collectible for large territories, so it is not probable that in the near future they will be available for Guatemala, since they require a huge amount of time and money to evaluate. Mora and Vahrson shows a lower precision, but starts from common data; basically a good DTM is needed and this requires less effort and no in situ analysis. Finally, the scale should be carefully considered: although these methods can be ideally applied to an entire country, the authors suggest choosing a proper area to investigate. The Italian case study has the right extension for methods application; since the results are good we can also expect good reliability on a small scale; in other cases, such as Guatemala for example, the low resolution of input data force one to move to a bigger scale, so the resulting map will be more likely to indicate hazard in extensive zones that will require further investigations. It is important to remark that all the models use approximation to describe a complex physical reality and to describe spatial variance of parameters, so they produce results with a limited reliability; however, these approximations are absolutely acceptable in large-scale evaluation, as in the cases analyzed in this paper, and can give a good idea of relative risk in an area.

6 Conclusions

This work demonstrates how it is possible to improve safety from landslide hazards, building thematic maps for large territories in a relative quick way. The objective of the authors was to focus attention on the critical situation of the Central American region and the lack of an effective approach to civil protection in that area. The methods applied here prove to give a precious indication about the vulnerability and can be used to rationalize the efforts and the scarce economical resources only in the high risk areas, being an effective DSS for a critical situation, improving effort effectiveness and the response to emergencies. Input data available for any situation will be an active part of the DSS; together with the space scale of results that is needed and the situation in which these models should operate they will determine when it is better to apply Mora and Vahrson and when SINMAP is to be preferred. Nowadays, input data for Guatemala are not as precise as they need to be for the correct application of models, but getting a new DEM with improved resolution is an operation with a reasonable price; it would certainly be more economic than a long in situ campaign. The models themselves can operate a good zoning operation when starting from high quality data and are suitable for the requested work. The Mora and Vahrson method requires more easily collectable data than SINMAP and gives slightly coarser results, but the evidence shows a good benefit-cost ratio. It is important to remember that the Mora and Vahrson method was developed and calibrated on the Central American situation and probably would have also given better results in the Guatemalan case study than in the Italian one, provided that it is possible to get the right data. So, the tools to improve safety and resilience to shallow landslides in Central America exist and are effective, and the authors hope that efforts will be made to improve safety conditions in those critical regions.

References

[1] Courela P.; Civil Protection and EU-Latin American Relations

[2] Goodwin C.N., Tarboton D.G., SINMAP User's Manual, A stability index approach to terrain stability hazard mapping

[3] Mora, S. and Vahrson, W. (1993). Macrozonation Methodology for Landslide in TC4 Manual for Zonation on Seismic Geotechnical Hazards.

[4] Hazard Determination, Bull. Intl. Ass. Eng. Geology, in press. Chisolm R.; The effect of rainfall on landslides in Guatemala during hurricane Stan, Report for Environmental and Water Resources Engineering, University of Texas at Austin
[5] Geopetrol SA Estudio hidro-geologico para lòa implementacion de un sistema de monitoreo y alerta ante deslizamentos en asentaminetos urbanos del departemento de Guatemala, Centro America pp 15-25
[6] Guatemala. Ministerio de Comunicaciones, Transporte y Obras Públicas. INSIVUMEH. Sección de Hidrología Aplicada, Inventario de los principales deslizamientos ocurridos en la República de Guatemala.

Section 4
Sediment transport and debris flow monitoring and analysis

Special session organised by
Daniele De Wrachien,
Gian Battista Bischetti,
Francesco Gentile & Luca Mao

Erosion and sediment transport modelling in Northern Puglia watersheds

F. Gentile, T. Bisantino & G. Trisorio Liuzzi
PROGESA Department, University of Bari, Italy

Abstract

In the Puglia region (southern Italy) heavy storms trigger high suspended sediment transport in water courses, accelerate soil and nutrients loss, and adversely affect biodiversity. In addition, high siltation in reservoirs reduces the water-holding capacity and creates severe problems of water availability for agriculture, which still plays an important role in the local economy. In the study area, suspended load data derive from continuous monitoring in the Carapelle torrent (2007-2008) and from hand-sampling in the Salsola sub-catchment of the Candelaro torrent (1970-1984). Recorded data of total streamflow are also available for both torrents. The high temporal resolution data were used to analyze the sediment transport dynamics and to evaluate the predictive accuracy of the Annualized AGricultural Non-point Source (AnnAGNPS) pollution model at the event scale. The historical data were used to test the reliability of the model for long-term periods and to compare the performances of medium and small size watersheds.
Keywords: soil erosion, sediment transport, AnnAGNPS model, continuous monitoring.

1 Introduction

During the past four decades, different simulation models have been developed to estimate surface runoff, sediment, nutrient and pollutant transport processes. The widely used water quality models include ANSWERS, Beasley *et al.* [1], CREAMS, Knisel [2], GLEAMS, Leonard *et al.* [3], AnnAGNPS, Bingner and Theurer [4], and SWAT, Arnold *et al.* [5]. Among these models the Annualized Agricultural Non-Point Source AnnAGNPS pollution model has a structure that balances complexity and parameterization as it uses empirical and quasi-

WIT Transactions on Engineering Sciences, Vol 67, © 2010 WIT Press
www.witpress.com, ISSN 1743-3533 (on-line)
doi:10.2495/DEB100171

physically based algorithms to predict runoff volume, peak flow rate, sediment and nutrient yield. AnnAGNPS was developed for simulation in ungaged agricultural watersheds, with the purpose of evaluating the influence of non-point source pollution on surface water and groundwater quality.

The model computes runoff using the SCS Curve Number method, which was originally developed for agricultural sites as an infiltration loss model. Using theoretical arguments it is possible to apply the SCS-CN method for hydrologic simulation to any basin (Mishra and Singh [6]). The method has several advantages over others because it is a simple conceptual method and is well supported by empirical data. AnnAGNPS has been calibrated, validated, and applied for runoff and sediment yield losses from watersheds in different geographic locations, conditions and management practices.

Applying AnnAGNPS and ANSWERS models (Walling *et al.* [7]) compared, firstly, observed and predicted runoff and sediment output data for individual storm events monitored at the basin outlets and, secondly, information on the spatial pattern of soil redistribution within the catchments derived from 137Cs measurements. The results obtained indicate that catchment outputs simulated by both models are reasonably consistent with the recorded values, although the AGNPS model appears to provide closer agreement between observed and predicted values. Comparison of the catchment sediment delivery ratios and the pattern of soil redistribution in individual fields predicted by the models with equivalent information derived from 137Cs measurements indicates that the AGNPS model provides more meaningful predictions of erosion and sediment yield than the ANSWERS model.

AnnAGNPS and SWAT models were calibrated in Red Rock Creek watershed and validated in Goose Creek watershed, both sub-watersheds of the Cheney Lake watershed. Forty-five months (1997-2000) of monthly measured flow and water quality data were used to evaluate the two models that performed well for surface flow and sediment yield (Parajuli *et al.* [8]).

AnnAGNPS reliability was assessed in the Mississippi Delta MSEA watershed (Yuan *et al.* [9]). Using no calibrated parameters, the underestimation of runoff for extreme events was observed, although the relationship between simulated and observed data on an event basis was significant (R^2 = 0.9). In contrast, the lower R^2 of 0.5 for event comparison of predicted and observed sediment yields demonstrated that the model was not best suited for short–term individual event sediment prediction. This may be due to the use of the Revised Universal Soil Loss Equation (RUSLE) within AnnAGNPS, and of parameters derived from long–term average annual soil loss estimates. The agreement between monthly average predicted and observed sediment yield had an R^2 of 0.7. Three–year predicted total runoff was 89% of the observed, and three–year predicted total sediment yield was 104% of the observed.

In this paper, the model is applied to the 506 km^2 Carapelle watershed and the 43 km^2 Salsola sub-catchment, which are predominantly agricultural with major crops of durum wheat. Flow regime is torrential and floods event are mainly associated with intensive, short-term rainfall. Soil erosion that affects these basins is an important indicator of soil productivity, while sediment yield influences water quality in agriculture. The sediment transport is mainly characterised by suspended materials.

The major objectives of this work are: to evaluate the ability of AnnAGNPS in simulating runoff and sediment yield at event scale in calibrated and validation modes; to compare the model estimation of runoff at event scale with the long-term simulation; to evaluate the model for runoff and sediment yield comparing the response of a medium size watersheds to that of a small size one in the same area.

2 Material and methods

2.1 The AnnAGNPS model

AnnAGNPS is a continuous simulation watershed-scale model developed on the single-event model AGNPS. AnnAGNPS simulates quantities of surface water, sediment, nutrients, and pesticides leaving the land areas and their subsequent travel through the watershed. AnnAGNPS divides the watershed into homogenous drainage areas, which are then integrated together by simulated rivers and streams, routing the runoff and pollutants from each area downstream. The hydrology of the model is based on a simple water balance approach that considers runoff, evapotranspiration and percolation, maintaining a water budget for the 2-layer soil system. The following equation is used to determine soil moisture for each time step in a day:

$$SM_{t+1} = SM_t + \frac{WI_t - Q_t - PERC_t - ET_t - Q_{lat} - Q_{tile}}{Z}$$

where:

SM_t = moisture content for each soil layer at the beginning of time period (fraction),
SM_{t+1} = moisture content for each soil layer at the end of time period (fraction),
WI_t = water input, consisting of precipitation or snowmelt plus irrigation water (mm),
Q_t = surface runoff (mm),
$PERC_t$ = percolation of water out of each soil layer (mm),
ET_t = potential evapotranspiration (mm),
Q_{lat} = subsurface lateral flow (mm),
Q_{tile} = tile drainage flow (mm),
Z = thickness of soil layer (mm),
t = the time period.

The influx is a function of effective rainfall less any retention required to wet the surface and fill depressions and any runoff determined by the day's runoff curve number. The initial infiltration into the control volume is predicted by the runoff curve number and is assumed to be a function of saturated flow into the control volume via worm holes and vertical cracks as well as flow through the interstices.

The flux out is the sum of the day's: (a) soil moisture leaving the bottom of the control volume; (b) lateral or quick return flow which would include tile drainage if present; (c) evapotranspiration which is calculated according to the FAO procedure.

The soil moisture movement within the control volume is determined using unsaturated flow equations for vertical percolation through the soil interstices. Percolation occurs at the rate of the hydraulic conductivity corresponding to the soil moisture content, calculated according to the Brooks-Corey equation.

Each day a daily runoff curve number is calculated based upon the given RCN II and the available soil moisture. Available soil moisture is the water content between the wilting point and field capacity. The daily runoff curve number is computed according to an exponential available soil moisture relationship where AMC I is at the wilting point (0%) and AMC II is half way between the wilting point and field capacity (50%). Tile drains, if present, affect the available soil moisture and its drainage can be a major source of quick return flow that adds directly to the recession leg of the runoff hydrograph.

Water that leaves the bottom of the control volume continues through the vadose zone and eventually becomes the major source of groundwater.

Soil erosion is determined using the Revised Universal Soil Loss Equation (RUSLE). Sediment yield is computed using the HUSLE equation (Theurer and Clarke [10]):

$$S_y = 0.22\, V_r^{0.68} q_p^{0.95}\ KLSCP$$

where S_y is the sediment yield (t/ha), V_r = surface runoff volume (mm), q_p = peak rate of surface runoff (mm/s), K,L,S,C,P are RUSLE factors. All three variables (S_y, V_r, and q_p) are based on unit area; i.e., divided by their drainage area.

Sediment transport is estimated using the Einstein deposition equation with Bagnold transport capacity.

Special components are included to handle concentrated sources of nutrients (feedlots and point sources), concentrated sediment sources (gullies), and added water (irrigation). Output is expressed on an event basis for selected stream reaches and as source accounting (contribution to outlet) from land or reach components over the simulation period. The model can be used to evaluate best management practices (BMPs).

2.2 Study area

The Candelaro and Carapelle torrents originate in the Apennine mountains and cross the Tavoliere flood plain before flowing into the Adriatic sea, fig.1. The watersheds are characterised by clayey-sandy Plio-Pleistocene sediments in the alluvial fan and by flyschoid formations in the mountainous areas, which are subject to erosion. The plain and the low hilly areas are mainly used for cultivation (85%) of durum wheat, high diversity of vegetables and olive groves, whereas forests and pasture prevail in the higher slopes. The climate is typically Mediterranean, with rainfalls ranging from 450 to 800 mm/year and average temperatures ranging from 10 to 16 °C.

Table 1: Main characteristics of the Carapelle at Ordona bridge and Salsola at Casanova watersheds.

		Carapelle at Ordona bridge	Salsola at Casanova
Watershed area	km^2	506.2	44.1
Maximum altitude	m a.s.l	1075.0	1025.0
Average altitude	m a.s.l.	466.0	432.0
Minimum altitude	m a.s.l	120.0	189.0
Main channel length	km	52.2	17.0
Main channel slope	%	1.8	4.9
Mean watershed slope	%	8.2	6.1

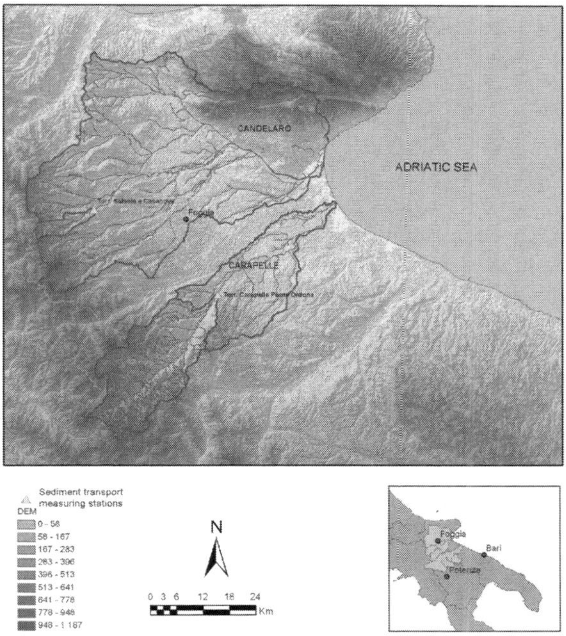

Figure 1: The Carapelle and Candelaro watersheds and relative sub-watersheds with the mouth at the sediment transport stations.

The erosion processes (rill and gully erosion) that affect the watersheds are mainly located on the hillslopes. During heavy storms these processes trigger high-suspended sediment transport, accelerate soil and nutrients loss, pollution in water courses and adversely affect biodiversity. Deforestation has increased the phenomenon of instability, landslides and unstable slopes, consequently runoff events cause high rates of sediment transport. Additionally high siltation in reservoirs reduces water-holding capacity and creates severe problems of water availability for agriculture and urban use. Typically rural farmers have problems related to water shortage.

Data sources for the model inputs included the NASA Digital Elevation Model for topography, the ACLA2 regional project (Caliandro *et al.* [11]), for soil data and the CORINE Land Cover cropland data layer.

2.3 Streamflow and suspended sediment data

Streamflow data for both study watersheds were available from the Italian Hydrological Service (IHS) gauging stations. Baseflow separation is required for numerous widely used hydrological and erosive models, such as AnnAGNPS, and it must be considered also in monthly models (Mouelhi *et al.* [12]). In this work the following filtering algorithms (Eckhardt [13]), for separating baseflow from total streamflow was used:

$$b_k = \frac{(1 - BFI_{\max})ab_{k-1} + (1-a)BFI_{\max}Q_t}{1 - aBFI_{\max}}$$

where b_k is the base flow at time step k; b_{k-1} is the base flow at the previous time step; Q_k is the measured total flow; BFI_{max} is a constant that can be interpreted as the maximum baseflow index; a filtering coefficient. The coefficient BFI_{max} or base flow index gives the long-term mean ratio of base flow to total runoff. BFI_{max} was calculated using the hydrograph recession curve analysis and the optimization module developed by Kyoung *et al.* [14].

The spatial distribution of rainfall data were assessed using the Thiessen weighting procedure for the closest rain gauges to the watersheds.

Continuous sediment load data derived from the monitoring station set up in the Carapelle torrent at Ordona-Castelluccio dei Sauri bridge. The station is equipped with an infrared optical probe (Hach-Lange SOLITAX Hs-line), chosen in view of its capability to measure high solid concentration and to reduce watery medium and light interferences. The probe measures suspended sediment coupling backscattering and nephelometric photodetectors. The probe was preliminary tested in laboratory using mixtures of varying granulometric concentration to evaluate its functional capacity and to assess the effects of the different solid fractions on the measurements. The instrument was field calibrated during the flood periods 2007-2009 (Gentile *et al.* [15]). Figure 2 shows the relation between sensor output and concentration as a result of previous calibration.

Afterwards, the instrument was tested in the field through a calibration stage and the verification of the instrument housing. The most relevant flood events were then considered and the suspended sediment concentration, monitored at half-hourly scale, was plotted versus discharge to analyze the sediment transport dynamic. The recorded discharges (Q) and sediment concentrations (SSCs) for the period 2007-2008 are reported in figures 3-4.

Monthly total streamflow are also available for the Carapelle watershed for the period 1986-1996, while yearly sediment load and monthly total streamflow data derive from IHS measurements for the Salsola sub-catchment (1970-1984).

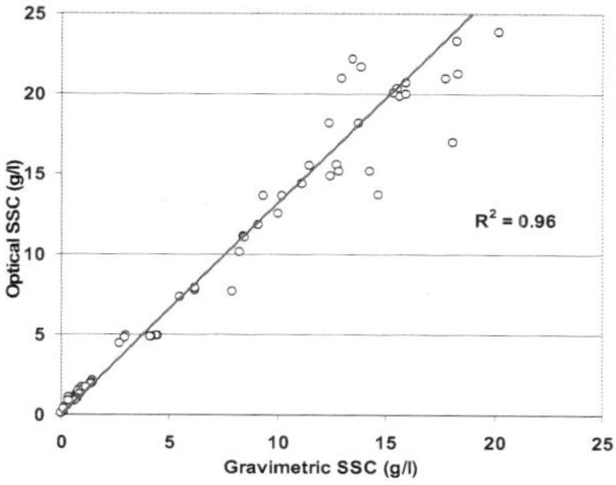

Figure 2: Relationship between sensor output and concentration.

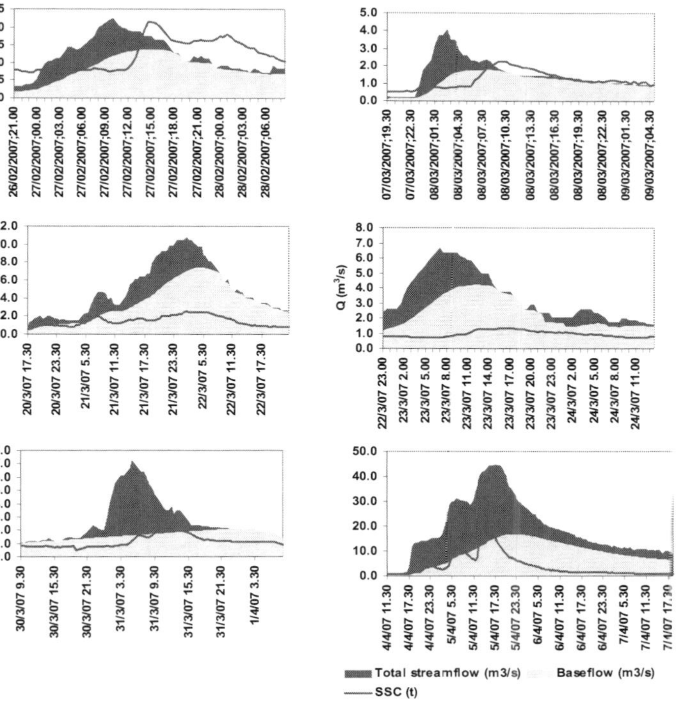

Figure 3: Flood events monitored in 2007 and used for the calibration of the AnnAGNPS model.

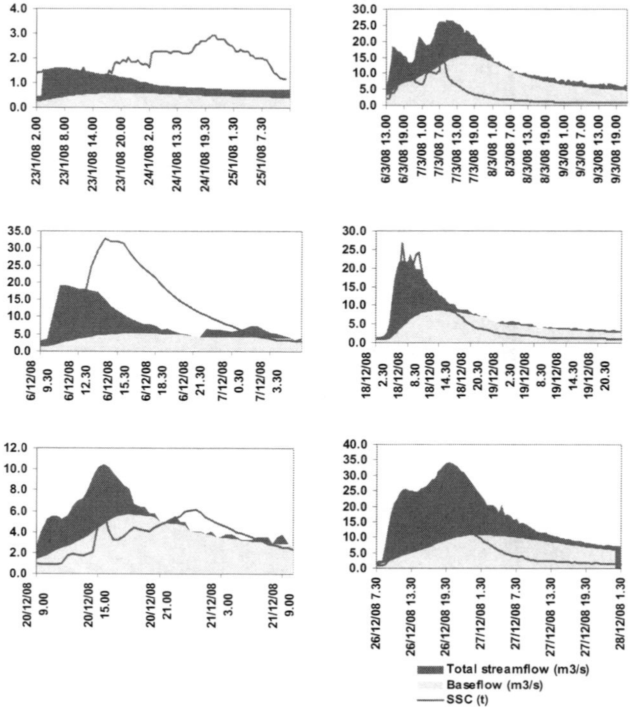

Figure 4: Flood events monitored in 2008 and used for the validation of the AnnAGNPS model.

Table 2: Initial curve number values.

Cover type	Initial curve numbers for hydrologic soil groups			
	A	B	C	D
Cropland	72	81	88	91
Fallow	76	85	90	98
Rangeland	35	56	70	77
Forest	43	65	76	82
Pasture	49	69	79	84
Urban	89	92	94	95

3 Results

3.1 Model preprocessing

The model preprocessing regarded the definition of the Curve Number data (tab.2) for each defined field type (cropland, fallow, rangeland, forest, pasture and urban) and for each Hydrologic Soil Group (A, B, C, D). The CN values were estimated for different land uses of the watersheds for both the watersheds and the weighted CNs for different cells of the agricultural watersheds were calculated.

The rainfall erosivity factor R was estimated based on the mean monthly precipitation of the period 1979-1999 according to Ferro *et al.* [16]. The value of R after calculations is 960.70 MJ mm ha^{-1} $year^{-1}$.

The Lal and Elliot [17] equation was used to estimate the soil erodibility factor K. Eight types of soils were identified, whose erodibility factors are shown in table 3. Soil types were used to evaluate the hydraulic soil properties such as the saturated hydraulic conductivity, field capacity and wilting point (Saxton and Rawls [18]).

The Crop data required by the model regard the units harvested, surface and subsurface decomposition, crop residue, root mass, canopy cover, management scheduling and operation. The management schedule of wheat was assumed (tab. 4) as cereals represent 84% of the total surface. Croplands contribute with relative low soil erosion. One factor contributing to the erosion of croplands is the amount annual disturbance. In this simulation, three disturbances have been chosen (plough, harvest and semi-deep drill).

The crop management factor C for each period was calculated based on land use, canopy cover, surface cover and surface roughness and soil moisture conditions. The P-factor was supposed to be equal to 1 since no management operation to reduce erosion has been considered.

Table 3: Soil erodibility factors K estimated using the soils' physical properties and organic matter content of the European Soil Data Centre (ESDAC).

Soil structure	**K factor (t h MJ^{-1} mm^{-1})**
Clay	0.033
Sandy clay	0.043
Loam	0.03
Clay Loam	0.042
Silty clay	0.0346
Silty-clay-loam	0.0271
Loamy-sand-clay	0.038
Sandy-loam	0.016

Table 4: Management scheduling and operation for wheat.

Crop	**Event date**	**Management schedule**	**Curve number**	**Management Operation**
Wheat	06/01/01	Harvest grain	Fallow	Residue added to surface
				Current crop harvested
				Call in a new crop growth
	09/01/01	Plough	Fallow	Soil disturbed
				Crop planting
	09/20/01	Begin crop growth	Cropland	Call in a new crop growth
	12/15/01	Semi-deep drill	Cropland	Soil surface disturbed

3.2 Model calibration and validation in the Carapelle watershed at event scale

Hourly streamflow and suspended sediment concentrations were used to calibrate and validate the model at event scale in the Carapelle watershed. A total of 6 events were used for model calibration, while 6 events were used in model validation.

Before the model calibration a preliminary sensitivity analysis (Chouaib [19]), for the most common parameters of the model (storm type, R, K, C and P factors of USLE equation, CN curve number and MN Manning's roughness coefficient) was carried out. The results pointed out that the CN coefficient is the most sensitive parameter as it controls runoff volumes while the storm type mainly influences peak discharge. CN was subjected to calibration varying the retention factor. After the calibration of CN values, Manning's coefficients were adjusted to fit peak discharge and sediment load.

The comparison of measured and calculated values of peak discharge, runoff volume and total suspended load during the events is shown in table 5. Simulated data were evaluated using statistical indexes, the coefficient of determination (R^2) and the Nash–Sutcliffe Efficiency Index (NSE). Correlation and agreement between observed and simulated peak flow, runoff and sediment load in the calibration phase is excellent. In the validation phase the model shows good performances for runoff, peak flow and sediment load.

Table 5: Measured and predicted peak discharge, runoff volume and sediment load for calibration and validation periods.

	Event	Peak discharge (m^3/s)		Runoff Volume (m^3)		Sediment load (t)	
		Obs	**Sim**	**Obs**	**Sim**	**Obs**	**Sim**
Calibration	27_2_07	1.1	1.3	94593	50807	202	586
	7_3_07	2.8	2.4	106362	92315	228	1048
	21_3_07	4.7	4.3	271983	230529	2178	2100
	23_3_07	3.3	0.2	135277	7669	452	43
	31_3_07	5.7	11.1	189558	443321	583	7820
	5_4_07	30.8	31.3	2896506	2826167	36124	28463
R^2		*0.94*		*0.99*		*0.91*	
NSE		*0.94*		*0.99*		*0.89*	
Validation	24_1_08	1.3	0.2	112969	7528	358	80
	6_3_08	14.3	38.5	1339256	2044547	18232	37310
	6_12_08	16.9	11.5	351810	429511	11336	11104
	18_12_08	18.4	5.0	508572	20781	13065	471
	20_12_08	5.8	3.0	145847	8175	1997	135
	26_12_08	25.7	32.9	1573264	1719026	25503	19841
R^2		*0.78*		*0.86*		*0.74*	
NSE		*0.54*		*0.6*		*0.65*	

3.3 Model application at the Salsola and Carapelle watersheds in the long-term

Calibrated parameters were used for the application at Carapelle (period 1986-1996) and Salsola (1970-1984) watersheds in order to test the model for the long-term. Comparison between simulated and observed runoff data in both the watersheds is reported in figures 5-6. Statistical parameters, that were calculated for each simulation (R^2=0.6, NSE=0.7 for the Carapelle and R^2=0.6, NSE=0.7 for the Salsola), point out that simulated runoff are in good agreement when compared with the observed watersheds response. Yearly sediment loads were calculated from complete monthly series. Sediment load modelling indicates simulated values are consistent with the recorded ones, fig. 7. This means that the modelling process is efficient for small and medium size watersheds when adequate meteorological, soil and crop data are available.

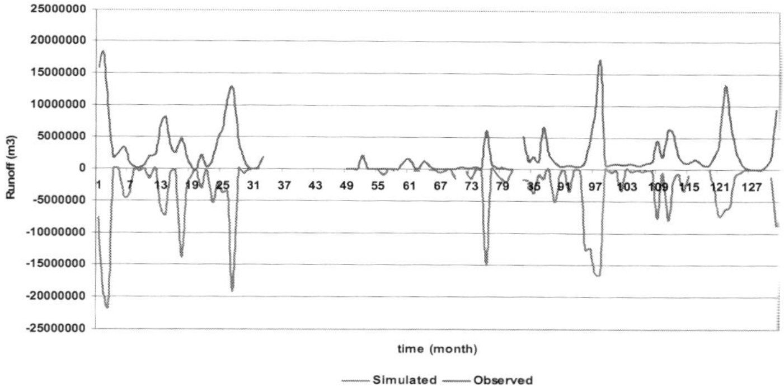

Figure 5: Runoff prediction of the AnnAGNPS model for the Carapelle watershed. In the upper part of the graph there are the observed runoffs, in the lower part the simulated ones.

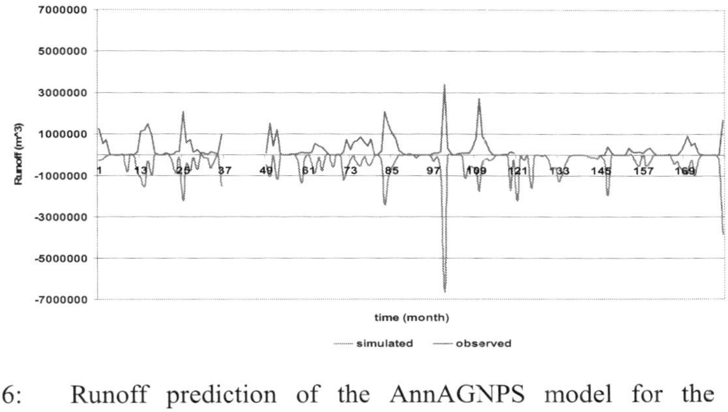

Figure 6: Runoff prediction of the AnnAGNPS model for the Salsola watershed. In the upper part of the graph there are the observed runoffs, in the lower the simulated ones.

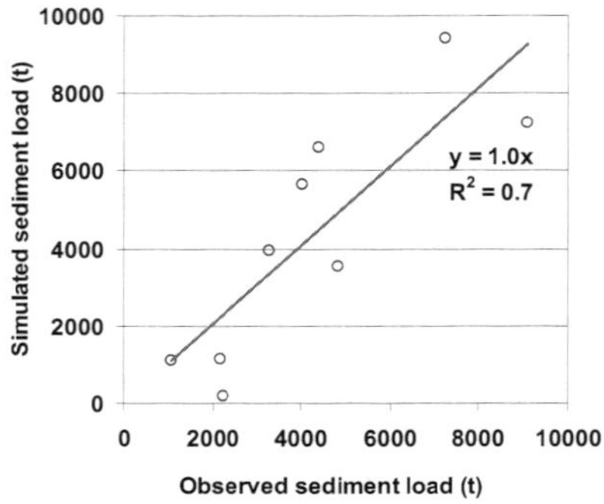

Figure 7: Yearly sediment load prediction of AnnAGNPS for the Salsola watershed.

4 Conclusions

In this paper the results of sediment transport modelling in Northern Apulia torrents are reported.

AnnAGNPS is a continuous simulation watershed-scale model for peak flow, runoff, sediment load and pollutants prediction. The model is very sensitive to storm type, CN values and to rooting density, surface residue and crown canopy cover associated with the C factor, so the application in uncalibrated mode can determine high average errors in predictions.

The AnnAGNPS model was calibrated and validated in the Carapelle watershed using twelve flood events. Very good performances were obtained by the model in simulating peak flow, runoff and sediment load which is mainly triggered by the field operations and by precipitation events.

Calibrated parameters were used to evaluate the model in the long-term using ten years of runoff data. Good correlation and agreement between simulated and observed data at both time scales allows one to use the model for studying sediment transport dynamics and for watershed management and prediction in ungaged basins having the same characteristics.

The long-term application at Carapelle watershed was compared to that of Salsola sub-catchment in order to test the influence of different spatial scales in modelling. The model has predicted the runoff volume within the range of good accuracy and this indicates that the SCS curve number method used in the AnnAGNPS model is suitable for runoff volume prediction. Simulated sediment loads values were compared with measurements showing a good agreement in the general pattern.

References

[1] Beasley, D.B., Huggins, L.F. & E.J. Monke, *ANSWERS: a model for watershed planning*. Trans. ASAE 23 (4), pp. 938–944, 1980.

[2] Knisel, W.G., CREAMS: A Fieldscale Model for Chemical, Runoff, and Erosion from Agricultural Management Systems. USDA, Science and Education Administration, Conservation Report No. 26, Washington, D.C., 1980.

[3] Leonard, R.A., Knisel, W.G., & Still, D.A., *GLEAMS: Groundwater Loading Effects of Agricultural Management Systems*. Transactions of ASAE, vol. 30, pp. 1403-1418, 1987.

[4] Bingner, R. L. & Theurer F.D. *AGNPS Web Site*, 2009. http://www.ars.usda.gov/Research/docs.htm?docid=5199

[5] Arnold, G., Srinavasan, R., Muttiah, R.S., & Williams, J.R., *Large Area Hydrologic Modeling and Assessment. Part I. Model Development,* Journal of the American Water Resources Association, vol. 34, pp. 73-89, 1998.

[6] Mishra, S. K. & Singh, V. P., *Soil Conservation Service Curve Number (SCS-CN) Methodology.* Series: Water Science and Technology Library, Vol.42, pp.536, Hardcover 2003.

[7] Walling, D.E., He, Q. & Whelan, P. A., Using 137Cs measurements to validate the application of the AGNPS and ANSWERS erosion and sediment yield models in two small Devon catchments. Soil and Tillage Research, Volume 69, Issues 1-2, pp. 27-432003.

[8] Parajuli, P. B., Nelson, N. O., Frees, L. D. & Mankin, K. R., *Comparison of AnnAGNPS and SWAT model simulation results in USDA-CEAP agricultural watersheds in south-central Kansas.* Hydrological Processes, 23(5), pp. 748-763.

[9] Yuan, Y., Bingner, R. L. & Rebich, R. A., *Evaluation of AnnAGNPS on Mississippi Delta MSEA Watersheds*. Trans. of the ASAE, Vol. 44(5): 1183-1190, 2001.

[10] Theurer, F.D. & Clarke, C.D., *Wash load component for sediment yield modeling.* In Proceedings of the Fifth Federal Interagency Sedimentation Conference, March 18-21, 1991, Las Vegas, NV: Subcommittee on Sedimentation of the Interagency Advisory Committee on Water Data, Vol. 1, pp. 7-1 to 7-8.

[11] Caliandro, A., Lamaddalena, N., Stellati, M. & Seduto, P., *Caratterizzazione agroecologica della Regione Puglia. In Funzione della potenzialità produttiva: Progetto Acla 2*. Puglia, Bari, 2005.

[12] Mouelhi, S., Michel, C., Perrin, C., & Andreassian, V., *Stepwise development of a two-parameter monthly water balance model*, Journal of Hydrology, 318, pp. 200–21, 2006.

[13] Eckhardt, K., *How to construct recursive digital filters for baseflow separation.* Hydrological Processes 19, 507-515, 2005.

[14] Kyoung, J. L., Jong-Gun, K., Bernie, E., Ji-Hong, J., Younshik, P., Yong-Chul, S., Sung-Gu, H., Ki-Sung, K., Joongdae, C., & Dong Sun, Y., *Development of Optimization Module in the WHAT System for Accurate Hydrograph Analysis and Model Application*. ASAE Annual Meeting. www.asabe.org. 2007.

[15] Gentile, F., Bisantino, T., Corbino, R., Milillo, F., Romano, G., & Trisorio Liuzzi G. *Monitoring and analysis of suspended sediment transport*

dynamics in the Carapelle torrent (southern Italy). Catena 80, pp. 1-8, 2009.

[16] Ferro, V., Porto, P., & Yu, B., A comparative study of rainfall erosivity estimation for southern Italy and southeastern Australia, Hydrol. Sci. – Journal-des Sciences Hydrologiques 44, pp. 3–23, 1999.

[17] Lal, R., & Elliot, W., *Erodibility and erosivity.* In: Lal R. (Ed.), Soil Erosion Research Methods. Soil and Water Conservation Society and St. Lucie Press. USA. pp. 181–208, 1994.

[18] Saxton, K.E., & Rawls, W.J., *Soil water characteristic estimates by texture and organic matter for hydrologic solutions*, Soil Sci. Soc. Am. J., 70, 1569-1578, 2006.

[19] Chouaib, W., *Evaluation of Annualized Agricultural Non Point Source model (AnnAGNPS) for runoff, peak flow and sediment yield estimations in the Carapelle watershed, Apuglia Region (Southern Italy).* Master of Science Thesis on Land and Water Resources Management: Irrigated Agriculture. International Centre for Advanced Mediterranean Agronomic Studies (CIHEAM-Bari Italy), 2009.

Restoration of a degraded torrential stream by means of a flood control system: the case of Arroyo del Partido stream (Spain)

J. A. Mintegui Aguirre, J. C. Robredo Sánchez,
C. De Gonzalo Aranoa & P. Huelin Rueda
*Escuela Técnica Superior de Ingenieros de Montes,
Universidad Politécnica de Madrid, Spain*

Abstract

The Arroyo del Partido is a small torrential stream (307.67 km^2), draining into the El Rocío marsh, in the Doñana National Park (Spain).The DNP was declared a UNESCO Heritage site (1994), since it has an important role for bird migration between Europe and Africa. The last reach (7 km) of the stream was channelized in 1981, thereby the flowing conditions during floods changed. This fact caused the formation of a large alluvial fan over the wetland surface during the period 1982-2003, covering an area of 4.31 km^2, with a volume of 4.48 × 10^6 m^3.

In order to contain the advance of the fan into the marsh a flood control system was used. This control system consists of two check dams, one of them placed on the channelized reach, and a second one placed on a different stream, which receives the surplus flows of the Arroyo del Partido during floods. Between both streams a flood plain was reserved for sediments deposition and flood attenuation purposes. Reconstruction works were carried out in the summer of 2006. During the period 2007-09, works were monitored and adjusted on the basis of their behaviour in flood events, with the objective of restoring the former hydrological regime before the channelization.

Keyword: major events, flood protection, hydrologic correction, ecological restoration.

1 Introduction

Originally the Arroyo del Partido presented the classical scheme of a torrential stream (catchment area, gorge, alluvial fan and discharge channel). Then, during

WIT Transactions on Engineering Sciences, Vol 67, © 2010 WIT Press
www.witpress.com, ISSN 1743-3533 (on-line)
doi:10.2495/DEB100181

major floods, surplus flow was expanded over the fan surface before reaching the Doñana National Park (DNP) marshes, generating the former alluvial fan. After these flood events, the laminated and devoid of sediments flow was concentrated within several drainage channels discharging to the marsh with a uniform flow regime. The fan surface was cultivated only during the years in which the winter flood occurred. The irrigation plan Almonte-Marismas transformed this area for crop purposes in 1981, channelling the last 7 km of the stream before the marshes. This transformation forced the concentration of the flood into the channel, and thus the former alluvial fan behaves as an extension of the gorge, which has caused the formation of a new alluvial fan at the end of the new channel, inside the DNP marshes. Data pertaining to the evolution of the new fan surface and volume, up to July 2003, is presented in Table 1.

Table 1: Surface and volume increment estimation of the new alluvial fan over the DNP marsh between 1956 and 2003.

Year	**Surface (m^2)**	**Volume (m^3)**
1956	0	0
1982	152.241	5.338
1985	292.142	19.657
1993	1.042.800	250.460
1996	2.035.211	954.017
1997	3.325.798	2.547.588
1998	3.814.778	3.733.575
2000	3.929.143	3.899.220
2001	3.979.757	3.973.527
2002	4.195.430	4.297.024
2003	4.314.186	4.479.898

The analysis of that situation drove the following proposal. In order to recover the former hydrological regime of the Arroyo del Partido it is necessary to: 1) construct a check dam for soil erosion control at the beginning of the old alluvial fan, about 6 km upstream of the inflow into DNP marshes and 2) permit the flooding of the adjacent surface during major flood events, downstream of the check dam cross section, to recover the function of the former floodplain. The works on the hydraulic and hydrologic restoration of the Arroyo del Partido were conducted in 2006, designated as Action Number 3 (A3) of the Project Doñana 2005, which is a major project on Water Regeneration of the DNP Marshes. Due to its great environmental impact, an Executive Committee was created, and a Scientific Committee for its monitoring.

2 Objective

Since the old alluvial fan permitted the flood expansion prior to the DNP marshes, thereby promoting the flood lamination and sediments deposition, A3 is aimed at achieving the same effects. On this basis, a Flood Control System was implemented to regulate the flow draining into the marshes. This implies the need for a new flood-lamination-sedimentation area over the old alluvial fan, or at least a part of it.

Hereby we analyze the performance of the works carried out under the A3 on the last reach of the Arroyo del Partido stream during the period 2006-09; so that to adopt if necessary the appropriate adjustments to the works to ensure over time the hydraulic and hydrologic restoration of the degraded reach and its environment. This text discusses: a) the Arroyo del Partido Flood Control System project, b) the way in which the field works were conducted at the focus site, c) the actual possibilities to generate the new flood-lamination-sedimentation plain, and d) how the stream flood should be discharged into the DNP marshes, such that the stream contours are not altered by sedimentary processes.

3 Methods and site description: works and measures adopted under the A3 of Project Doñana 2005

The area where the Arroyo del Partido Flood Control System was located is shown in Figure 1. In the aerial photograph the Arroyo del Partido drains along the left-hand side of the image, and the Cañada del Pinar stream, having its own catchment, drains along the right-hand side. At the bottom of the image, from west to east, it the road that connects El Rocío and Villamanrique villages can be distinguished. The concerned area presents a slope towards the Cañada del Pinar stream, the reason why this last stream acted in the past as an additional drainage

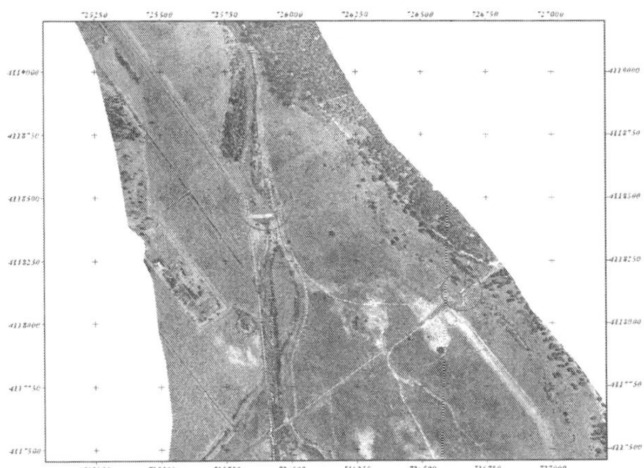

Figure 1: Aerial photograph (2008) showing the flood-lamination-sedimentation area recovered with the A3. The Arroyo del Partido drains along the left-hand side of the image and the Cañada del Pinar stream along the right-hand side. The road connecting El Rocío and Villamanrique runs along the bottom of the image. CD1 and CD2 are highlighted with circles. Triangles represent topographic landmarks, used for the levelling of the studied area.

channel for the floods coming from the Arroyo del Partido stream. The works implemented in order to achieve the aforementioned control system consisted of: a) the construction of a check dam on the El Partido channel at the beginning of the former alluvial fan (Check Dam 1 or CD1), controlling channel erosion processes upstream from its location, and distributing the stream flow between the Arroyo del Partido channel and the flood-lamination-sedimentation area, b) the construction of a second check dam on the Cañada del Pinar stream (Check Dam 2 or CD2), regulating the flood level in the floodplain, and c) the space available for flooding, flood lamination and sediments deposition, placed between both streams, and devoted in A3 for this purpose.

The CD1 height is 2.6 m above the current streambed. Originally the frontal spillway was 2.0 m high, 70.0 m long, and 2.0 m width, situated at an average altitude of 16.7 m. Thirty one weep holes, with 0.3 m in diameter, crossed the structure from the upstream to the downstream side, draining the water volume stored behind the structure. The weep holes were arranged in two rows (of 15 weep holes each) and a single bottom drain (Figure 2). The lateral spillway was placed immediately upstream of the structure along the left riverbank and coated with a rocky breakwater and a cyclopean concrete lining in order to prevent erosion when the diverged discharge flows into the flood-lamination-sedimentation area. The lateral spillway has a length of 140 m and its altitude ranges between 16.70 and 16.77 m.

The CD2 is 1.5 m high and was built immediately upstream of the bridge over the Cañada del Pinar stream, on the road connecting El Rocío and Villamanrique (Figure 3). Its spillway is 29.5 m long, 2.0 m high and 3.0 m width, with an average altitude of 14.62 m. The structure is crossed by 12 weep holes, 0.3 m in diameter, draining the volume stored behind the check dam, arranged in two rows (of 6 weep holes each).

The area comprised between both streams belonged to the former alluvial fan, before its channelization (1981). Accordingly, the A3 should recover its function. It is composed of two different areas:

a) The 57 hectares, triangular-shaped, surface upstream of the El Rocío-Villamanrique road. A geometric levelling of its perimeter was performed, since it was the first part of the plain being flooded, in order to estimate the behaviour

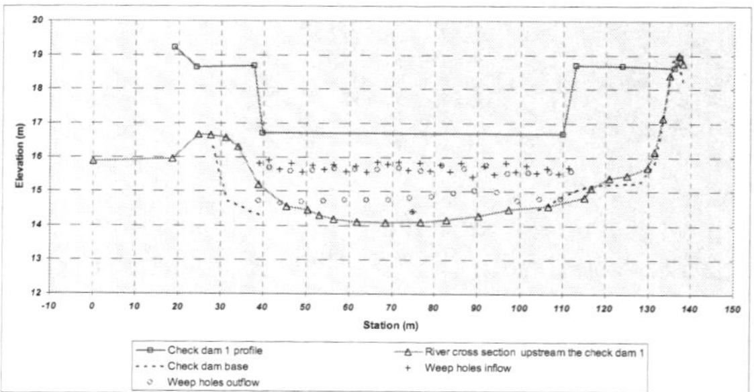

Figure 2: CD1 elevation view (left) in 2006 and photograph (right) in July 2007.

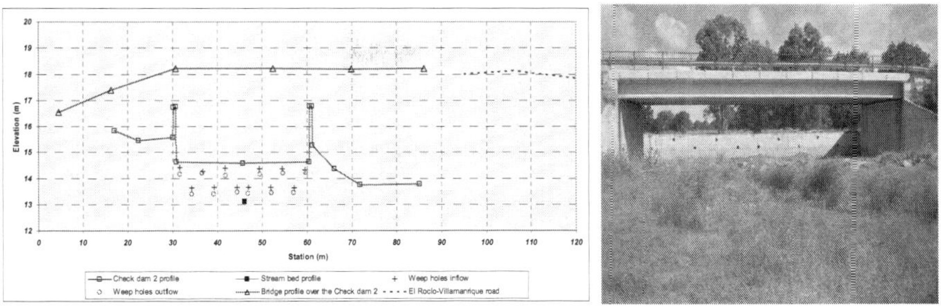

Figure 3: CD2 elevation view (left) in 2006 and photograph (right) in July 2007.

of the flood that occurred on July 19th, 2003. However, the greatest concern was focused on the geomorphologic response of the terrain surface of this area, when the Arroyo del Partido surplus flow comes from the lateral spillway of the CD1. For this reason, within the initial Flood Control System approach, we opted for a conservative flow allocation at the CD1. Using a design flow of a flood having a 100-year recurrence interval (358 $m^3 \cdot s^{-1}$), the flow would be distributed as follows: a 40% would discharge through the frontal spillway and weep holes, and the remaining 60% would flow through the lateral spillway.

b) The area placed downstream from the mentioned road, occupying 227 hectares, which can be flooded in the case of extreme floods. In such cases the road would operate as a discharge threshold of the stored water, flowing from the upstream surface (a) to the continuation of the floodplain area (b). A geometric levelling (October 18th 2008) of the road was conducted between both courses in order to establish the spill level of the road.

4 Results and discussion: behaviour of the works and regulations undertaken in the A3. Adjustments of the initial design

4.1 Stage-discharge curves of the El Partido flood control system at the Check Dam 1 and 2

The initial design of the CD1 (built in 2006) was tested, and it was found that the control system performed efficiently for major floods, but did not prevent minor flooding and problems related to sediments deposition in the eastern surroundings of El Rocío village, placed at the edge of the marshes. The stabilization of the new alluvial fan, formed over the marshes, was not solved either, because the flow reaching there was still too high. Moreover, the floods did not flood, as expected, the area devoted for that purpose. In these circumstances, as a step towards implementing the necessary adjustments of the CD1, the stage-discharge curves of the Flood Control System were determined at the CD1 and CD2. We only show here the stage-discharge curve at the CD1 (Figure 4, left chart), being the only one that suffered corrections.

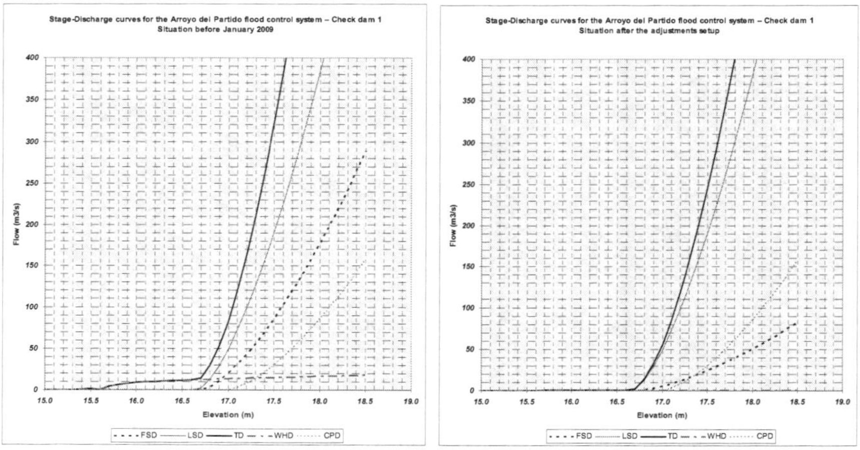

Figure 4: Flow allocation ($m^3 \cdot s^{-1}$) at the CD1 among the different discharge sections: left chart, with the initial design (2006), and right chart after the adjustments in January 2009. Frontal spillway discharge (FSD), Lateral spillway discharge (LSD), Total discharge (TD), Weep holes discharge (WHD), Cañada del Pinar discharge (CPD).

4.2 Adjustments of the Check Dam 1 and new stage-discharge curves of the Flood Control System

The adjustments of the CD1 were carried out after the summer of 2008, and were operational at the beginning of 2009. Their aim was the reduction of the flow crossing the CD1 to The Arroyo del Partido channel through its front face. With that purpose the following measures were implemented: a) to modify the initial frontal spillway section (2006), decreasing its draining surface, b) to cover the weep holes with gravel, delaying the discharge through them during floods.

Accounting these modifications, the new frontal overflow section became trapezoidal, 20 m long at the base and 24 m long at the crest, maintaining a height of 2 m (Figure 5). The new stage-discharge curves of the Flood Control System at the CD1, after the modifications are shown in Figure 4, right chart.

The effectiveness of these adjustments was tested by analyzing their behaviour during two flood events on February 1st and 5th of 2009 (Figure 6).

The results of the first flood were surprising. The pressure of the flow dragged the gravel blocking the weep holes through them, becoming those lasts operative again, as shown in the photograph in figure 5. The estimated flow distribution at the CD1in these events, is shown in the Table 2.

The photograph in figure 5 shows the situation of the CD1 at 13:00 h (local time) on February 2nd 2009, during the flood that started the previous day. At that time, the estimated hydrograph (figure 6) points out a flow rate of 15 $m^3 \cdot s^{-1}$. This value matches with the stage-discharge curve at the CD1 at the same time, as shown in the photograph in figure 5, where all the weep holes were draining at their maximum capacity and there is no discharge over the frontal spillway threshold.

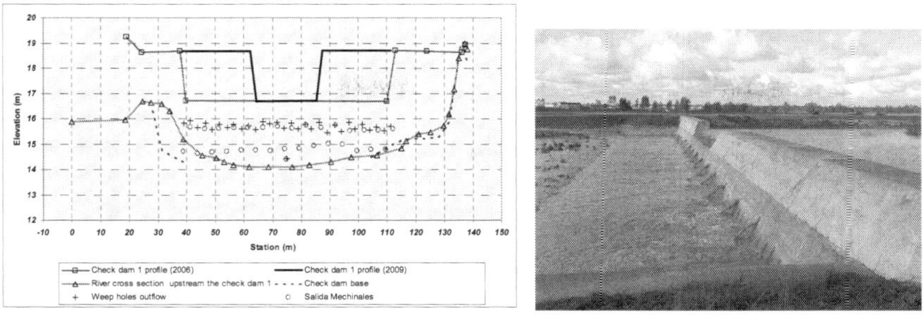

Figure 5: Left: CD1 elevation view with the final frontal spillway design. Right: Photograph of the CD1 at 13:00h on February 2, 2009, after the flood of the previous evening (image courtesy of C. Urdiales, DNP).

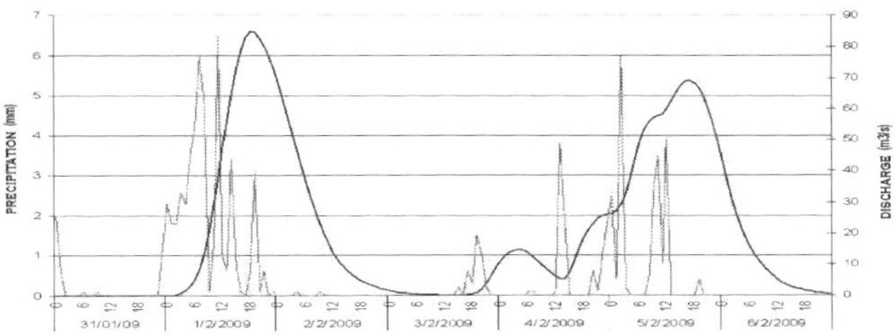

Figure 6: Precipitation at the Bollullos meteorological station (placed approximately at the centre of the Arroyo del Partido catchment) between January 31st and February 5th, 2009 (light grey line). Peak flow estimated at the CD1 upstream face on February 1st and 5th of 2009 (dark grey line).

Table 2: Flow distribution considering different discharge sections at the CD1 and the backwater effect at the beginning of the Cañada del Pinar stream during the floods of February 1st and 5th 2009.

CD1 scenario during the flood	Peak flow ($m^3 \cdot s^{-1}$)	Peak flow distribution through different spill elements ($m^3 \cdot s^{-1}$)			
		Weep holes	Frontal spillway	Lateral spillway	Backwater in Cañada del Pinar
Frontal spillway (2009) and operative weep holes.	70*1	15	6	49	-
	85*2	15	9	58	3

*1 Estimated flow allocation on February 5 2009. *2 Estimated flow allocation on February 1 2009.

Figure 7: Flood in the Arroyo del Partido stream at 13:10 h local time on February 5 2009. Left: flow distribution at the CD1. Right: the flood flows downstream over the flood-lamination-sedimentation plain.

The left photograph in Figure 7 shows the flood on February 5th 2009 at the CD1, at 13:10h (local time). As the hydrograph shows (Fig. 6), the stream flow at that moment was about 65 $m^3 \cdot s^{-1}$, where 20 $m^3 \cdot s^{-1}$ crossed the check dam front face (15 $m^3 \cdot s^{-1}$ through the weep holes and 5 $m^3 \cdot s^{-1}$ over the spillway threshold). The remaining flow, 45 $m^3 \cdot s^{-1}$, were diverged to the flood-lamination-sedimentation area over the lateral spillway, as shown in the same photograph, and ran downstream over the floodplain, as shown in the right photograph in Figure 7.

The analysis of these two floods allowed verifying the response of the drainage streambeds in the flood-lamination-sedimentation area when flooded. The analysis was possible due to the photographs taken while the area was flooded (Figure 7, right photograph); but more decisively due to the later topographical levelling, made on June 6-2009. With this survey, it was detected that the flood flow, discharged over the lateral spillway at the CD1, was adapted to the micro-topography of the plain, defined by the old drainage channels. The flood only caused little abrasion erosions at the base of the lateral spillway of the CD1. The profiles obtained for both floods are shown in Figure 8, together with the scheme of the Arroyo del Partido Flood Control System.

4.3 Drainage of El Partido floods into the DNP marshes, after being laminated and devoid of sediments at the area devoted for this purpose

Once the suitability of the flood-lamination-sedimentation area response was proved, it was analyzed the feasibility of a final stream discharge into the marshes in a uniform flow regime, in order to not alter the channel contours with sedimentary processes.

With this purpose a previous investigation (Urdiales, 1996) was taken into account. It was based on the (still conserved) profiles of the former Arroyo del Partido channel, running close the El Rocío surroundings. According to it, before the Arroyo del Partido channelization in 1981, the stream concentred the flow in a drainage channel with a very specific morphology, locally known as *caño*, to

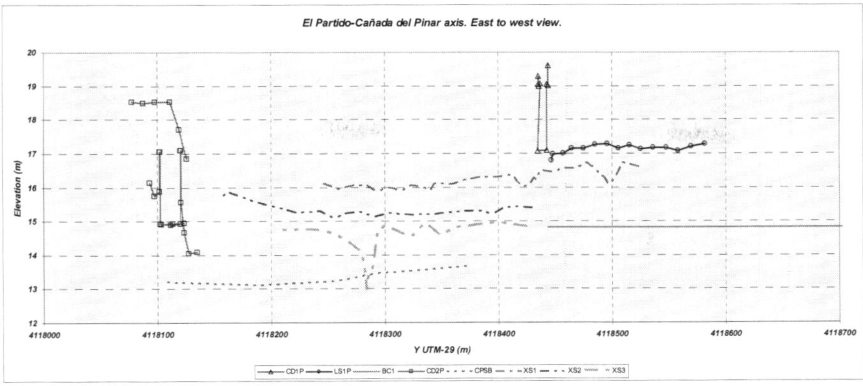

Figure 8: Elevation view (from East to West) of the elements composing the Flood Control System, and flood response on February 1 and 5 2009. From left to right: CD2 profile (CD2P); Cañada del Pinar streambed profile (CPSB); cross sections defined by floods in the flood-lamination-sedimentation area, surveyed on June 6 2009 (XS1, XS2, XS3); CD1 profile (CD1P); CD1 lateral spillway profile (LS1P); CD1 backwater curve (BC1).

flow towards the marsh. With the current Monitoring, it was detected that after the CD1 was constructed, which reduced substantially the flood flow reaching El Rocío village, a recovery of the Arroyo del Partido stream has been produced compared to the period between 1982 and 2006, especially regarding the recovery of the morphology of the former drainage system (caño), which was already lost. This fact ratified the modifications made to the CD1 to achieve the hydraulic and hydrologic restoration of the last reach of the Arroyo del Partido stream. Several cross sections along the last reach of the stream prior to the marsh inflow were levelled for confirmation. One of them is shown in Figure 9. The graph on the left corresponds to a 0.30 $m^3 \cdot s^{-1}$ flow. It was checked that in that situation the new alluvial fan was stabilized. The maximum evacuation capacity of this caño was estimated in 4 $m^3 \cdot s^{-1}$ (right graph in Figure 9), and this value was considered in the A3 for the CD1frontal spillway rectification. As a first approach, it was estimated that this flow corresponds to the bankfull stage of the channel, i.e. the evacuation capacity related with the dominant channel forming flow, linked with its compensation slope (García Nájera, 1943). This last was determined, by means of direct field survey, in 0.0016 $m \cdot m^{-1}$. Accordingly, it was considered that for flow rates under 4 $m^3 \cdot s^{-1}$, the sediment supply to the marsh remains beneath a moderated limit, since the flood remains into the drainage channel, and the sediment transport is essentially restricted to suspended load. Above this value, the best option is to permit the flood expansion across both overbanks. For this purpose it is necessary to ensure enough space, moving away any protection wall to the outer limits of the predictable flood expansion. In this way, the shear stress is insufficient to erode the banks and sweep away sediments into de marsh.

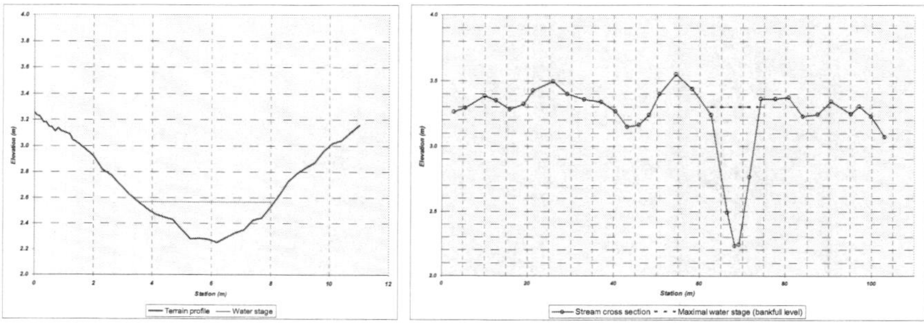

Figure 9: Left: Arroyo del Partido cross section, on the last reach prior to the marsh inflow, gauged on October 17 2008. Right: the same cross section, assuming a 4 $m^3 \cdot s^{-1}$ flow (bankfull level).

5 Conclusions

1) The conducted Monitoring consists on an approximation procedure to estimate the capacity of the A3 to move upstream the starting point of the alluvial fan to its ancient location and to reactivate its discharge channels to route the flood to the DNP marsh. This prevents from flooding and aggradation at the eastern side of El Rocío village, and maintains the current outlines of the marsh.
2) Regarding specifically the results related with the protection of the eastern side of the El Rocío village, the A3 estimates that: a) while the flood flow upstream the CD1 is lower than 20 $m^3 \cdot s^{-1}$ the situation will be under control. b) If the flood flow at the same location is lower than 100 $m^3 \cdot s^{-1}$, the direct discharge to the stream channel will reach a theoretical final value not greater than 5 $m^3 \cdot s^{-1}$, in the case that all the weep holes remain blocked (except for the bottom drain). This value, which could be seldom exceeded, ensures that the situation is also controlled. The flood discharge on February 1st and 5th 2009 was estimated over 20 $m^3 \cdot s^{-1}$ and no flooding or aggradation problems were detected on the eastern area of El Rocio village. Nevertheless, a planning is needed to ensure that the flood in the stream behaves as expected for a caño in this environment, i.e. concentring 4 $m^3 \cdot s^{-1}$ in the self drainage channel and expanding the surplus flow across both overbanks. So that it is necessary to remove any obstacle to the flow, preventing from the formation of new discharge channels westwards. This would promote undesirable situations around El Rocío village (in any case, not as serious as in the past). c) If the flood flow upstream the CD1 is greater than 100 $m^3 \cdot s^{-1}$, it has to be considered that the estimated flow getting to the surroundings of El Rocío village are reduced to less than the 15% of the reference rate before the construction and modification of the CD1 (2009). It is also noted that these flow rates, despite being high, have an order of magnitude that permits a planning and have a lower frequency. Another positive effect is that the A3 guarantees the ecological flow regime in the Arroyo del Partido stream, which contributes to its hydro-ecological recovery.

3) Regarding the functionality of the Cañada del Pinar stream, as an additional drainage of the flood-lamination-sedimentation area devoted in A3, it is noted that it was operative in the past. In this sense the floods on February 1st and 5th 2009 reactivated it without causing remarkable problems. As far as the Cañada del Pinar morphology remains as a caño, it is likely that the floods through it may behave in a similar way to the flow diverged from the lateral spillway of the CD1 over the flood-lamination-sedimentation area during the aforementioned floods. Therefore it is likely that the capacity of the Cañada del Pinar stream to drain the flood-lamination-sedimentation surface towards the marsh is compatible with the conservation of the current outline of the marsh, keeping away the threat of a new alluvial fan over the marsh.

Acknowledgements

We would like to thank the Scientific Commission of the Project Doñana 2005, the Works Director of the Action number 3, B. Bayán, and the Conservation Area Manager of the DNP, C. Urdiales, for their permanent collaboration in this Monitoring

References

[1] García Nájera J.M. (1943) *Principios de Hidráulica Torrencial y su Aplicación a la Corrección de Torrentes*. Instituto Forestal de Investigaciones y Experiencias. Madrid. 297 pp.

[2] Mintegui J. A., Robredo J. C., Sendra P. J. (2003) *Avenidas torrenciales en el arroyo del Partido y su incidencia en la marisma del Parque Nacional de Doñana,* pp. 373, Naturaleza y Parques Nacionales, Serie Técnica, Organismo Autónomo Parques Nacionales, Madrid.

[3] Mintegui J. A., Lenzi M. A., Robredo J. C., Mao L. (2006) *Movilización versus estabilización de los sedimentos en cursos sometidos a la dinámica torrencial,* pp. 143, Naturaleza y Parques Nacionales, Serie Técnica, Organismo Autónomo Parques Nacionales, Madrid.

[4] Mintegui J. A., Robredo J. C., De Gonzalo C. Huelin P. (2009) Seguimiento de la Actuación núm. 3 del Proyecto Doñana 2005, 139 pp., E. T. S. de Ingenieros de Montes, Departamento Ingeniería Forestal, Universidad Politécnica de Madrid.

[5] Saura J.; Bayán B.; Casas J.; Ruiz de Larramendi A.; Urdiales C. (2001) *Documento marco para el desarrollo del Proyecto Doñana 2005, Regeneración hídrica de las cuencas y cauces vertientes a las marismas del Parque Nacional de Doñana*, pp. 201, Ministerio de Medio Ambiente.

[6] Urdiales C. (1996) Informe Cambios recientes en el tramo final del arroyo del Partido, Parque Nacional de Doñana, Organismo Autónomo Parques Nacionales, Ministerio de Medio Ambiente.

The effects of large wood elements during an extreme flood in a small tropical basin of Costa Rica

L. Mao[1] & F. Comiti[2]
[1]*Department of Geography, University of Hull, UK*
[2]*Faculty of Science and Technology, Free University of Bozen-Bolzano, Italy*

Abstract

In-channel large wood elements (LW) enter the river system as a consequence of natural processes independent of channel activity (i.e. windthrow, large landslides, extended wildfire, disease outbreaks), or as a result of river processes (bank erosion, avulsions). LW exert a substantial beneficial influence on river morphology and freshwater biodiversity, but on the other hand may increase flood hazards in densely inhabited floodplains. This is due to a variety of processes, such as flow surges following collapse of temporary wood dams, strengthening of debris flows, local bed scour and local bank erosion, and clogging of culverts and bridges. However, despite a general consensus among populations and river managers about the hazards related to floating wood during floods, few studies have addressed the real contribution of LW to flood damages, as well as its sources and effects on flood dynamics. This paper presents a post-event analysis in the Río Portalón, a small tropical basin in the Pacific coast of Costa Rica, which experienced an extreme flood in September 2005. On that occasion, floating wood apparently had a considerable role in flood dynamics and in causing damages to the nearby village. Interviews with local people and a simple field survey were used to shed light on event dynamics, and in particular on role played by LW. Finally, a range of options for the management of riparian vegetation in tropical areas are discussed.

Keywords: riparian vegetation, large wood elements, tropical basin, flood risk, river management.

WIT Transactions on Engineering Sciences, Vol 67, © 2010 WIT Press
www.witpress.com, ISSN 1743-3533 (on-line)
doi:10.2495/DEB100191

1 Introduction

Large wood elements (LW, i.e. woody elements > 10 cm in diameter and > 1 m in length) have a substantial beneficial influence on river morphology [16, 25], hydraulics and sediment transport [11, 28] and freshwater biodiversity [2, 7]. LW can enter the river system as a consequence of processes independent of channel activity (i.e. windthrow, large landslides, extended wildfire, disease outbreaks), or as a result of river processes (bank erosion, avulsions). LW sources are thus subject to a very relevant spatio-temporal variability due to the activation of the associated processes, so that wood input rates can vary considerably for streams in the same region and even for different reaches of the same river [23]. The amount of in-channel large wood is therefore strongly connected to both the availability of trees in the riparian and basin area, and to the magnitude and frequency of input and output processes. Beside the positive effects on stream ecosystems, in-channel wood may also increase flood hazards. This can be caused by a variety of processes, such as flow surges following the collapse of temporary wood dams, inclusion and strengthening of debris flows, local bed scour and local bank erosion, and clogging of culverts and bridges [13]. Nonetheless, these potential hazards are strongly dependent on the degree of human presence within the catchment (frequency and type of road crossings, proximity and density of human infrastructures adjacent to the channels).

Despite a general consensus among populations and river managers about the hazards related to floating wood during floods, few studies have addressed the real contribution of LW to flood damages, or its sources and effects on flood dynamics. Comiti et al. [8], analysing a flash flood event in a mountain basin of the Slovenian Alps, concluded that the transfer of wood elements from the headwaters to the main channel was mostly associated with debris flows, and other LW sources were represented by forested floodplains in aggrading reaches. Entrained LW were then trapped by bridges, which were under-designed even for water and sediment fluxes alone. Therefore, in contrast to the generalized feeling that LW were indeed the main "culprit" for those damages, Comiti et al. [8] argued that LW possibly increased the final damages by exacerbating an already critical situation.

This paper presents a simplified post-event analysis in the Río Portalón, a small tropical basin in the Pacific coast of Costa Rica, which experienced an extreme flood in September 2005. On that occasion, floating wood apparently had a considerable role in flood dynamics and in causing damages to the nearby village. Interviews with local people and a basic field survey were used to shed light on the event dynamics, and in particular on role played by LW. Finally, a range of options for the management of riparian vegetation in tropical areas will be discussed.

2 Study site and description of the flood event

The Río Portalón is a small (17 km^2) mountain basin (maximum elevation of 760 m a.s.l) on the west coast range of Costa Rica, 40 km south of Quepos (Fig. 1).

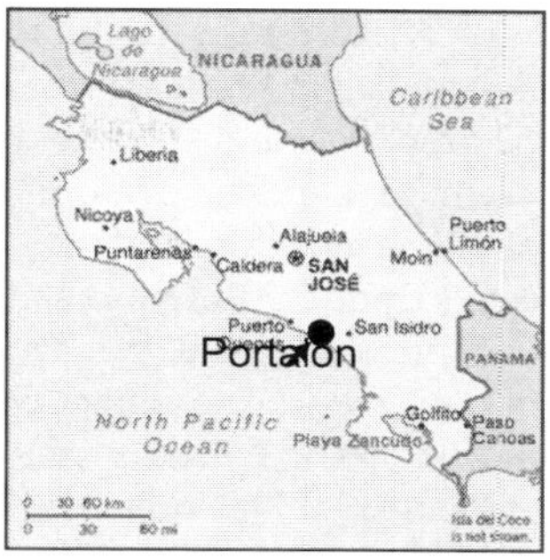

Figure 1: Location of the Río Portalón basin.

Figure 2: Multiple landslides triggered by the September 2005 event in the Río Guabas. The general condition of the Río Portalón has been reported to be similar (photo taken from [3]).

The river exits its confined segment at the hamlet of Portalón and then flows for 4 km towards the ocean on the narrow coastal plain.

The basin features steep hillslopes, with widespread landslides and debris flows. Soils are generally very thick, ranging from 3 to 10 m in depth, and the underneath bedrock is mainly a turbiditic sedimentary rock of the Tertiary age [24]. In September 2005, the passage of the hurricane Rita generated diffuse flooding and mass wasting on the whole Central pacific area of Costa Rica, creating major economic and social impacts on the area [24]. In the Río Portalón basin, several landslides were triggered as a consequence of the intense precipitations (Fig. 2). The landslides supplied the main channel and the tributaries with large quantities of sediment and large mature trees.

Interviews to Portalón residents - carried out on February 2008 - indicate that, on September 23 2005, very intense rainfall started at 10 AM and continued until the late evening, eventually leading to a high flood peak. The event transported relevant quantities of sediments and driftwood, impacting heavily the Portalón village. Most of the people interviewed refer to the formation of a "wood plug" in the confined reach, which caused the flood surge once it collapsed. Witnesses also reported that most of the wood was transported during the peak of the flood. Furthermore, local people claimed a similar event in the Río Portalón occurred about 50 yr ago. Forty-three houses were completely destroyed and 300 people had to be rescued and relocated from the Portalón and other villages of the area. Furthermore, there were extensive damages to electric and water supply systems as well as the collapse of an important bridge on a national road, which prevented emergency vehicles from reaching the area for few days. A preliminary survey of the Comisión Nacional de Prevención de Riesgos y Atención de Emergencias (CNE) reported that the main channel along its lower reach (near the village), aggraded by 2 to 4 m [24].

3 Field-based estimation of the flood peak discharge

A field survey along the Río Portalón main channel was carried out on February 2008, in order to test the hypothesis that a dam-break surge was released by the collapse of a LW jam (the "wood plug" reported by the inhabitants). After an inspection of the entire channel, 7 reaches were identified (Fig. 3). Seven cross-sections were measured along with the longitudinal bed slope. High-water marks (i.e. deposited driftwood, scars on trees, gravel deposits) were used to estimate the peak stage for each cross-section (Fig. 4). Bed slope ranges between 0.015 and 0.05, whereas bankfull width varies between 16 m and 23 m.

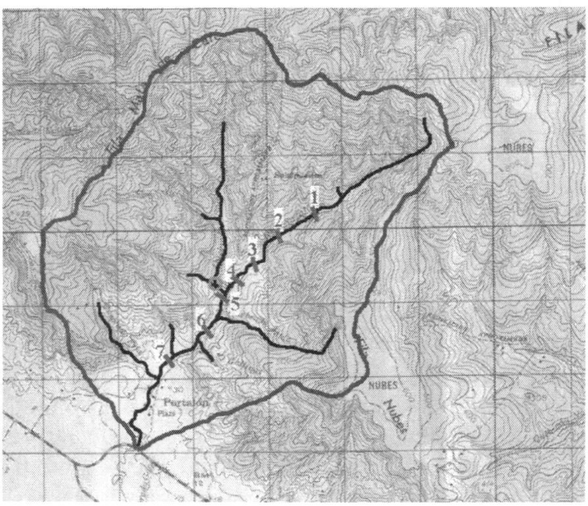

Figure 3: The Río Portalón basin and the location of the seven cross-sections where the peak flood has been estimated.

Figure 4: Field survey of cross-sections (left) and an example of flood mark (tree scar, right) Note the presence of very large boulders (D > 1.5 m) within the channel.

Simple methods have been used in order to estimate the maximum flood discharge based on the field evidences, due to the lack of any gauging station in the basin. The first method relies on the assumption that critical flow conditions ($Fr \approx 1$) were established along the reach during the flood, being critical conditions a limiting state in high-gradient streams featuring cascade and step pool morphologies [9, 10, 15, 20]. The critical velocity (V_c, in m s^{-1}) is calculated from the estimated mean depth (assumed to be critical, h_c, in m) and the acceleration due to gravity (g = 9.81 m s^{-2}) as:

$$V_c = \sqrt{gh_c} \tag{1}$$

The second method used in this study is Jarrett's [17] formula for the calculation of Manning's roughness coefficient. Jarrett developed an equation which relates channel roughness to bed slope (S) and hydraulic radius (R, in m). Reformulated in terms of mean velocity (V), the equation reads as follows:

$$V = 3.17R^{0.83}S^{0.12} \tag{2}$$

The third method is based on flow competence, i.e. the maximum clast size of the transported sediments. This was assumed to be the largest boulders still lying on the bed surface and that had been transported by the event. The intermediate axis diameters of the 10 largest boulders found around the cross-section were measured. The critical velocity required to move the averaged maximum clast diameter (D, in m) was calculated by the empirical formula of Costa [12]:

$$V = 5.2D^{0.49} \tag{3}$$

For all the methods, once the velocity is estimated, the peak discharge (Q, in m^3 s^{-1}) is obtained multiplying velocity by the cross-sectional wetted area (in m^2).

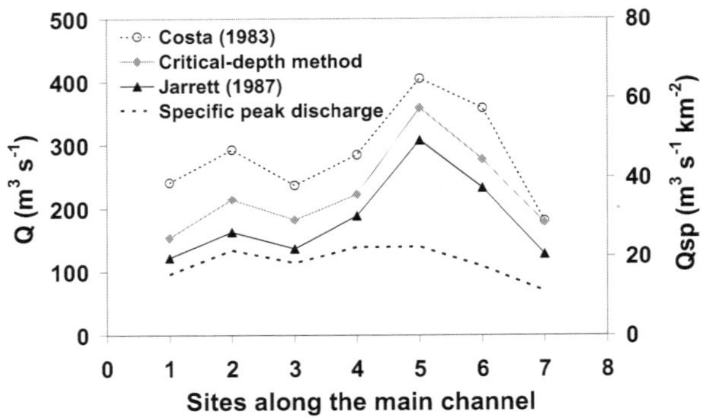

Figure 5: Estimated peak flood discharges (Q) along the Río Portalón main channel, from upstream (1) to downstream (7) cross-sections. The specific peak discharge (Q_{sp}) trend based on the critical-depth discharge calculated values is also showed.

The three mentioned methods used to estimate the peak discharge led to rather different results in absolute terms (Fig. 5). The method based on flow competence represents an upper estimate for peak discharges, whereas the Jarrett's formula provides values of Q which are, on average, lower by 8% than the critical-depth method which is considered to be the most reliable one. However, it is worth noting that the relative longitudinal trend showed by all methods is consistent.

4 Wood dynamics and effects during the flood event

Looking at Fig. 5, peak discharge appears to steadily increase downstream due to the larger catchment area down to section 4. Peak discharge then nearly doubles (from 220 to 360 $m^3 s^{-1}$) passing from section 4 to 5, i.e. just after the confluence with a right tributary, the Quebrada Salto, to abruptly decrease at downstream sections 6-7.

A field inspection revealed that the basin drained by the Quebrada Salto was heavily affected by landslides which supplied large amount of sediments and large tress to that tributary. The narrow width of the Quebrada Salto has likely led to the formation of temporary jams of logs and boulders, with a considerable capacity to store water and sediments behind them. Despite the lush vegetation which rapidly grew at the channel margins after the flood, some evidences of the presence of large log jams were observed in at least 2 sites along the Quebrada Salto. In particular, the presence of several wood pieces just in front of the confluence with small debris flow channel, along with immobile boulders and sediment accumulation upstream reveals the likely position of one of this log jam (Fig. 6), a few hundred meters upstream of the junction with the Río Portalón (circle on Fig. 3). However, it is worth noting the large specific peak discharge (~ 20 $m^3 s^{-1} km^{-2}$) characterizing the event already upstream of the Quebrada

Salto, thus suggesting that the extreme flood magnitude is most likely attributable to the very intense precipitations.

The collapse of such jam (about 4m-high) during the flood event may have caused a small surge wave transporting sediment and wood into the main channel, and the impulsive nature of the flood at its peak was indicated by local people during the interviews. However, the sharp increase in peak discharge at section 5 is nevertheless consistent with the augmented drainage area (Fig. 3). On the other hand, the sudden drop in the estimated peak discharges at section 6 hints to a rapid attenuation of the flood wave typical of small dam-breaks in mountain rivers (Fig. 5). As to the even lower discharge value at section 7, it can be attributed to the attenuation due to some overbank flows occurred on the left side of the channel between cross-section 6 and 7.

A quick survey of wood pieces along the analyzed segment of the Río Portalón revealed a low amount of LW volume, i.e. 7 m^3 per ha of channel area.

Figure 6: Boulders and wood pieces in one of the likely locations of the log jams.

Figure 7: Large trees growing at the margins of the Río Portalón main channel.

Wood elements were relatively large, with an average length and diameter of 6.2 m and 0.37 m, respectively. However very large trees are ready to be recruited at channel margins (Fig. 6).

Even though very little is known about volumes and dynamics of wood in tropical rivers, the measured wood quantities are comparable to what Gomi et al. [14] found on headwater channels of the Peninsular Malaysia. The limited amount of in-channel wood pieces is due to their relatively short residence time in such environments, because of the frequent flashy floods events with large transport capacity as well as for the high decay rates [6, 14]. However, recent investigation in several other streams of Costa Rica featuring lower bed slopes [6] showed that wood loading can be rather large even in tropical basins. The very limited quantity of wood lying within the Portalón channel contrasts with the huge amount of driftwood transported by the September 2005 flood and deposited on the floodplain at the Portalón village (Fig. 8). Field observations and reports by witnesses suggest that most of the large trees which destroyed the buildings and the bridge were recruited from the forested banks just upstream of the town, where bed aggradation led to channel avulsion. This is in agreement with the fact that long wood elements are unlikely to travel long distance in narrow channels [5, 19].

Figure 8: Wood elements lying on the flooded portion of the Portalón village (left) and accumulated on the collapsed bridge (right). The photos have been taken from the on-line version of the newspaper Al Dìa (26th September 2005).

5 Management options for large wood and riparian vegetation

Forest effectiveness in protecting hillslopes from soil erosion, landslides and debris flows for "ordinary" events is widely recognised [1, 27]. However, the forest cover may not prevent high magnitude geomorphic events from occurring, and sometimes mature plants removed and transported to the river network can increase the catastrophic effect and the hazard impacts of flood events due to wood jam collapses, reduced conveyance of critical cross-sections, and increased loading conditions on bridge piers. Such an adverse effect usually drives the decision of river managers to "clean-up" the river network from in-channel wood and to periodically cut riparian vegetation on floodplain. This practice is detrimental to stream ecology but on the other hand has also dubious effectiveness to reduce the hazards of high magnitude flood events. As showed by the Río Portalón case, wood stored in channels is only a little portion of the wood transported during extreme events, and the movement of in-channel logs is much reduced during low to moderate floods. Thus, for most period of time the in-channel wood lies relatively stable in the river system contributing to stabilize the bed, to limit the sediment transport and to create discontinuities that provide habitats for fishes and macroinvertebrates communities. Considerable amount of wood, which can create unstable jams, is recruited from slopes instabilities spread on the basin and eroded banks only during high magnitude floods. These infrequent floods are able to transport substantial loads of sediments and modify substantially the river bed due to erosion/deposition processes, which can represent the main threat even without any floating wood.

In the densely populated European and Japanese Alps, excessive wood transport is controlled by in-channel control works for wood retention. The choice of the most appropriate retention structure depends on channel width and on the amount of large wood potentially transportable by flood events [21]. Filter concrete check dams, rope net barriers and cable filter dams are the most effective structures [21]. To reduce the vulnerability of single in-channel infrastructure such as bridges, solutions as debris sweepers - vertically affixed to the upstream side of piers - might be an option, but their effectiveness under extreme flood conditions is yet to be demonstrated [4].

A more effective management option against flood risk – not only due to floating wood – would be to avoid any reconstruction or building of valuable structure in flood-prone areas. The adoption of a careful land use planning should of course be negotiated with local communities in order to achieve a prevalent social acceptance. As a compatible strategy for flood hazard reduction, wood recruitment and transport processes have been recently modelled [18] and considered in hazard mapping [22, 26]. However, these mapping methods usually require the availability of well calibrated probabilistic regional curves for flood-design estimation and detailed maps of landslide prone areas land use and forest vegetation typologies [22], and have been implemented experimentally only in Alpine areas.

6 Conclusions

In the Río Portalón, field surveys and interviews to local people indicate that large quantities of trees were supplied to the channel network during the 2005 flood event, likely creating some temporary wood dams. However, the flow surge associated to their collapse is very unlikely to have determined the flood itself, whose catastrophic consequences can be traced to the severity of the precipitation event. The reduced amount of wood lying on the channel contrasts with the huge quantity of trees transported during the flood. This suggests that the cleaning of wood from the river network is not an effective strategy in reducing flood hazards since wood jams in the upper part of the basin are created by trees delivered by landslides, and wood transported to the village are likely to be supplied by floodplains in the lower section of the stream. In-channel control works such as rope net barriers and cable filter dams are effective in capturing driftwood but are usually expensive and require high maintenance. It is thus preliminarily suggested that, in a situation like the study case, a more effective strategy option against flood risk would be to avoid any reconstruction within the floodplain, where the Portalón village was and still is located.

Acknowledgements

This research was funded by the EU Project INCO-CT-2004-510735 EPIC FORCE (Evidence-based policy for integrated control of forested river catchments in extreme rainfall and snowmelt). L. Mao is currently supported at the University of Hull by a Marie-Curie fellowship (EU 7th Framework Programme; PIEF-GA-2008-219294).

References

[1] Andreassian, V., Waters and forests: from historical controversy to scientific debate. *Journal of Hydrology*, 291, pp. 1-27, 2004.

[2] Benke, A.C., Wallace, J.B., Influence of wood on invertebrate communities in streams and rivers. American Fisheries Society Symposium, 37, pp. 149-177, 2003.

[3] Bonilla, E., Chaves, I., Esquivel, L., Madrigal, J., Mendez, J. & Sjobohm, L., *Casos frecuentes de Amenaza y Vulneraqbilidad*. Medidas de prevencion y mitigacion, Comisión Nacional de Prevención de Riesgos y Atención de Emergencias, Costa Rica, 2006.

[4] Bradley, J.B., Richard, D.L., Bahner, C.D., Debris control structure, evaluation and countermeasures. *Hydraulic Engineering Circular*, 9, U.S. D.T., pp. 179, 2005.

[5] Braudrick, C.A., Grant, G.E., Transport and deposition of large woody debris in streams: a flume experiment. *Geomorphology*, 41, pp. 263-283, 2001.

[6] Cadol, D., Wohl, E.E., Goode, J.R., Jaeger, K.L., Wood distribution in neotropical forested headwater streams of La Selva, Costa Rica. *Earth Surface Processes and Landforms*, 34, pp. 1198–1215, 2009.

[7] Collier, K.J., Halliday J.N., Macroinvertebrate-wood associations during decay of plantation pine in New Zealand pumice-bed streams: Stable habitat or trophic subsidy? *Journal of the North American Benthological Society*, 19, l, pp. 94-111, 2000.

[8] Comiti F., Mao L., Preciso E., Picco L., Marchi L., Borga M., Large wood and flash floods: evidence from the 2007 event in the Davča basin (Slovenia). In *Monitoring, Simulation, Prevention and Remediation of Dense and Debris Flows II*, eds. De Wrachien D., Brebbia C.A., Lenzi M.A., WIT Press, 60, pp. 173-182, 2008.

[9] Comiti F., Mao L., Wilcox A., Wohl E.E., Lenzi M.A., Field-derived relationships for flow velocity and resistance in high-gradient streams. *Journal of Hydrology*, 340(1-2), pp. 48-62, 2007.

[10] Comiti, F. Cadol, D., Wohl, E., Flow regimes, bed morphology, and flow resistance in self-formed step-pool channels. *Water Resources Research*, 45, W04424, doi:10.1029/2008WR007259, 2009.

[11] Comiti, F., Andreoli, A., Mao L., Lenzi M.A., Wood storage in three mountain streams of the Southern Andes and its hydro-morphological effects. *Earth Surface Processes and Landforms*, 33, pp. 244-262, 2008.

[12] Costa, J.E., Paleohydraulic reconstruction of flash-flood peaks from boulder deposit in the Colorado front range. *Geological society of American bulletin*, 94, pp. 986-1004, 1983.

[13] Diehl, T.H., Potential drift accumulation at bridges. U.S. Department of Transportation, Federal Highway transportation, FHWA-RD-97-028, 1997.

[14] Gomi, T., Sidle R.C., Noguchi, S., Negishi, J.N., Nik, A.R., Sasaki, S., Sediment and wood accumulations in humid tropical headwater streams: Effects of logging and riparian buffers. *Forest Ecology and Management*, 224, pp. 166-175, 2006.

[15] Grant, G.E., Critical flow constrains flow hydraulics in mobile-bed streams: A new hypothesis. *Water Resources Research*, 33(2), pp. 349-358, 1997.

[16] Gurnell, A.M., Piegay, H., Swanson, F.J., Gregory S.V., Large wood and fluvial processes. *Freshwater Biology*, 47, pp. 601-619, 2002.

[17] Jarrett, R.D., Errors in slope-area computations of peak discharges in mountain streams. *Journal of Hydrology*, 96, pp. 53-67, 1987.

[18] Lancaster, S.T., Hayes S.K., Grant, G.E., Modelling sediment and wood storage and dynamics in small mountainous watersheds. *Geomorphic Processes and Riverine Habitat*, ed. Dorava J.M. et al., Water Sci. Appl., AGU, Washington DC, 4, pp. 85-102, 2001.

[19] Mao L., Andreoli A., Comiti F., Lenzi M.A., Geomorphic effects of large wood jams on a sub-Antartic mountain stream. *River Research and Application*, 24(3), pp. 249-266, 2008.

[20] Mao L., Comiti F., Lenzi M.A., La resistenza al flusso in un torrente montano ad elevata pendenza con morfologia a step-pool. *Rivista di Ingegneria Agraria*, 3, pp. 29-38, 2006.

[21] Mao L., Comiti F., Andreoli A., Picco L., Lenzi M.A., Urciulo A., Iturraspe R., Iroumè A., Role and management of in-channel wood in relation to flood events in Southern Andes basins. In *Monitoring, Simulation, Prevention and Remediation of Dense and Debris Flows II*, eds. De Wrachien D., Brebbia C.A., Lenzi M.A., WIT Press, 60, pp. 207-216, 2008.

[22] Mazzorana, B., Zischg, A., Largiader, A., Hubl J., Hazard index maps for woody material recruitment and transport in alpine catchments. *Natural Hazards and Earth System Sciences*, 9, pp. 197-209, 2009.

[23] Meleason, M.A., Davies-Colley, R.J., Hall, J.M.J., Characterizing the variability of wood in streams: simulation modelling compared with multiple-reach surveys. *Earth Surface Processes and Landforms*, 32(8), pp. 1164-1173, 2007.

[24] Méndez Herrera, J.C., Esquivel, L., Inundaciones, deslizamientos, flujos de lodo y detritos: Sus efectos sociales y geológicos en las microcuencas de los ríos Guabas y Portalón (September, 2005). *IX Seminario Nacional de Geotecnia*, San José, Costa Rica, 14-18 November 2006.
[25] Montgomery, D.R., Collins, B.D., Buffington, K.M., Abbe T.B. Geomorphic effects of wood in rivers. In *The Ecology and Management of Wood in World Rivers*, eds. Gregory. S.V., Boyer K.L., Gurnell A.M., American Fisheries Society, Bethesda, MD, pp. 21-47, 2003.
[26] Rigon, E., Il legname in alveo nei torrenti alpini: *Analisi quantitativa e modellazione GIS*. PhD thesis, University of Padova, Italy, pp 178, 2009.
[27] Sidle, R.C. & Ochiai, H. *Landslides: Processes, Prediction, and Land Use.* Water Resources Monograph, AGU, Washington DC, 18, pp. 312, 2006.
[28] Wilcox, A., Nelson, J.M., Wohl E.E., Flow resistance dynamics in step-pool channels: 2. Partitioning between grain, spill, and woody debris resistance. *Water Resources Research*, 206, 42, W05419. DOI: 10.1029/2005WR004278, 2005.

Rheological properties and debris-flow modeling in a southern Italy watershed

T. Bisantino[1], P. Fischer[2], F. Gentile[1] & G. Trisorio Liuzzi[1]
[1]*PROGESA Department, University of Bari, Italy*
[2]*ETH Zurich, Institute of Food Science and Nutrition, Switzerland*

Abstract

The Pulsano watershed, located in the southern side of Gargano (Puglia-Italy), is subject to debris-flow phenomena originating from the weathering of the limestone rocks that constitute the geological parent material. The territory is poorly monitored and between intense and rare events, which transport large particle size materials, more frequent hyper-concentrated flows, which transport heterogeneous sediments, occur. On this basis the debris-flow risk assessment appears very complex, since it depends on both the available debris volume and the hydraulic characteristics of the flow. This work proposes a methodology of analysis and modeling of debris-flows that is able to integrate the results of the experimental investigations on materials with models simulating the triggering, propagation and deposition of the flow. For this purpose the rheological response of natural suspensions collected from the superficial deposits was investigated using a rheometric tool for large particle suspensions: the Ball Measuring System (BMS). The triggering and propagation of the debris-flow were simulated using the SHALSTAB and FLO-2D models in order to reproduce the 15 July 1972 event. Afterwards the influence of rheology on debris-flow depositional features was investigated.

Keywords: debris-flow, rheology, modeling.

1 Introduction

In recent years some debris-flow events occurred in the South-Gargano watersheds (Puglia Region-Southern Italy). These phenomena originate from the weathering of the limestone rocks that constitute the geological parent material

WIT Transactions on Engineering Sciences, Vol 67, © 2010 WIT Press
www.witpress.com, ISSN 1743-3533 (on-line)
doi:10.2495/DEB100201

of the promontory and are fostered by the inadequate protection of the degraded forests.

In this work an integrated approach for debris-flow analysis and modeling was set up to study the poorly documented event of July 1972, which occurred in the Pulsano watershed (Gentile *et al.* [1]). The methodology uses a physically-based model (SHALTAB, Montgomery and Dietrich [2]), which identifies the areas of potential shallow landsliding, and a two-dimensional debris-flow routing model (FLO-2D, O'Brien *et al.* [3]) that calculates depths, velocities and runout distances of the mixture.

The geo-mechanical and rheological properties of the materials involved in debris-flows were investigated as they represent key factors for physical and numerical modeling of the debris-flow dynamic. To define the geo-mechanical behavior of the soil a first batch constituted of ten soil samples was collected from the Pulsano watershed (Gentile *et al.* [4]). For the rheological experiments a second batch, constituted of four soil samples, was selected and analyzed using a rheometric system for fluids with large particle size: the ball measuring system (BMS) (Schatzmann *et al.* [5], Bisantino *et al.* [6]). The Bingham rheological model was fitted to the experimental data. The results of the experiments were analyzed with reference to the sediment concentration.

The output of debris-flow modeling was compared with field observations. Finally, the depositional features of the debris-flows were investigated varying the rheological parameters.

2 Materials and methods

2.1 Debris-flow modeling

The total debris volume triggered by heavy rains can be estimated using empirical formulas (Takei [7], Kronfellner-Kraus [8], PWRI [9], D'Agostino *et al.* [10], Marchi and Tecca [11], Marchi and D'Agostino [12], Tropeano and Turconi [13]), geomorphologic approaches (Hungr et al. [14]), or stability models such as SHALSTAB (Montgomery and Dietrich [2]), SINMAP (Pack *et al.* [15]), LISA (Hammond *et al.* [16]), and the one developed by Iverson [17].

The efficacy of the empirical formulas is limited by the variability of debris-flow phenomena and by the sediment availability (Rickenmann [18]). On the other hand the geomorphologic approach is influenced by the evaluation of the debris material that can be mobilized (Brochot *et al.* [19]). Physically-based modeling is an important tool for magnitude assessment as it considers the hydraulic, morphologic and geo-mechanical properties of materials during the event.

SHALSTAB is a coupled hydrologic and slope stability model that assumes steady state and saturated flow parallel to the slide surface. It states that shallow landslides are influenced by the topography, the convergence of subsurface runoff, the saturation degree of materials and the reduction of the shear stress in the granular mixture. The coupled model is represented by the following equation:

$$Q = \frac{Tsen\theta}{a / b}\left[\frac{c}{\rho_w gz\cos^2\theta\tan\phi} + \frac{\rho_s}{\rho_w}(1 - \frac{\tan\theta}{\tan\phi})\right]$$

where a is the drainage area, b is the outflow boundary length, θ is the hillslope angle, T is the transmissivity, Q is the effective rain, ρ_s is the soil density, c is the cohesion and ϕ is the friction angle.

The FLO-2D model was used to perform the runout modeling. It is based on the solution of the conservation of mass and momentum equations and uses a quadratic rheological model defining the flow pattern of the mixture both for a viscous fluid and for a turbulent/dispersive fluid. The resistant term of the Saint Venánt equations has a higher influence on reproducing debris-flows than the other terms (Arattano *et al.* [20]), so it should be adequately estimated. The rheological behavior of a debris-flow containing coarse particles cannot be assessed considering exclusively the contribution of the finer matrix (silt and clay) and thus neglecting the effects of direct grain contacts (Sosio *et al.* [21]).

2.2 Rheological parameters

The rheology of a suspension can be analyzed through measurements of shear stress τ at various shear rate $\dot{\gamma}$. These parameters have been investigated by several authors, for different debris mixtures, through laboratory systems such as parallel plates, concentric cylinder rheometers, slump test, inclined plane test and belt conveyor (Coussot and Piau [22], Malet *et al.* [23]).

Fine sediment mixtures at low concentrations show a Newtonian behavior whereas at higher concentrations they have a viscoplastic behavior represented by the Bingham (yielding with subsequent Newtonian flow), Herschel-Bulkley (yielding with subsequent power-law flow) models, bilinear (Locat [24]), (Newtonian flow at low shear stress and Bingham flow at high shear stress), quadratic (O'Brien *et al.* [3]). The last model describes the continuous flow regimes from viscous to turbulent/dispersive flow:

$$\tau = \tau_y + \eta\,\dot{\gamma} + C\,\dot{\gamma}^2$$

where τ is the shear stress, τ_Y the yield stress, $\dot{\gamma}$ the shear rate, η the dynamic viscosity and C is the inertial shear stress coefficient (Bagnold [25]). For mixtures having a mainly viscous behavior the quadratic term can be neglected and the model leads to the Bingham one. The viscosity η and the yield stress τ_y are expressed as an exponential function of the volumetric concentration of fine sediments:

$$\eta = \alpha_1 e^{\beta_1 C_v}$$

$$\tau_y = \alpha_2 e^{\beta_2 C_v}$$

where α_i and β_i are empirical coefficients defined in experiments.

In recent years, debris-flow materials containing large size particles have been examined. The results show that: the Herschel-Bulkley model can fit the data and that the exponent n and the ratio k/τ_c of the equation remain constant for suspensions obtained adding coarse particles to a large content of fine fraction (Coussot *et al.* [26]); for shear rates > 10 s^{-1} the Bingham model obtains better results than the Herschel-Bulkley one (Hübl and Steinwendtner [27]), meaning that the material under flow behaves more Newtonian than shear-thinning; at the same concentrations C_v gravel-mud mixtures having a larger content of fine sediments (d<1mm) have higher values of viscosity and yield stress (Wang and Jan [28]).

A recently developed rheometrical tool, the ball measuring system (BMS), has been used to analyze fine and large particle suspensions (Schatzmann *et al.* [29]). In the BMS the dragged measuring sphere can be considered as a particle on which the interactions with the fluid or the other particles are measured. A comparative analysis between the BMS, large-scale rheometers, and other commonly accepted methods was performed using debris-flow materials taken from Eastern Switzerland (Kaitna *et al.* [30]). The experiments, involving mixtures with different concentrations and particle size up to 5 mm, demonstrated that the results of the experiments obtained with the BMS are quite consistent with those obtained with large-scale rheometers, slump test, and inclined plate tests.

2.3 Study site and input data

The study was applied to map the flooded areas of the Pulsano torrent during the July 1972 storm event. The area extends from the Gargano promontory ($altitude_{max}$ = 880 m a.s.l.) down to the alluvial fan where the agricultural and industrial areas of Manfredonia spread, fig. 1.

The torrent has a watershed characterized by erosion processes on the hillslopes, failing stone walls, scouring and bank failure along the hydrographic network. The parent material of the watershed is mainly composed by dolomitic limestone (Triassic-Cretaceous era). Some field surveys and laboratory tests were carried out on the colluvial deposits to evaluate the geo-mechanical properties of the soil matrix that belongs to the class of loamy-sand materials, tab. 1.

The vegetal cover is characterized by natural grazing, prairies and broad-leaved woods at high altitudes; olive trees, crops and natural areas in the alluvial fan. The climate is typically Mediterranean, with rainfalls ranging from 450 to 780 mm/year and average temperature ranging from 10 to 16°C. On July 15, 1972 a catastrophic debris-flow inundated Manfredonia. Four valleys within the town turned into conveyance channels and the flow reached depths h > 1.25 m as mapped by Bissanti [31], fig. 2. The estimated return period of the rainfall is T_r > 200 years.

Four soil samples, collected from the Pulsano watershed, were selected for the rheological tests. The samples contain high amount of large particles (d < 4 mm) and some clasts with a diameter > 4 mm. The clasts create jamming

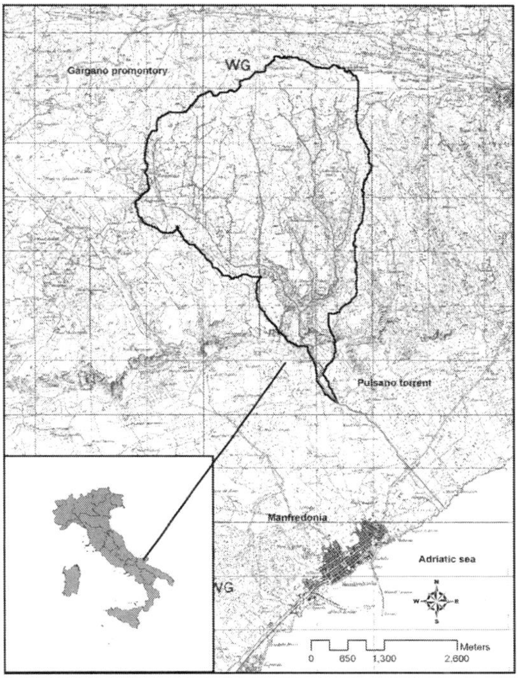

Figure 1: The Pulsano watershed.

Table 1: Main characteristics of the Pulsano watershed. Physical and mechanical properties of the soil.

Basin area A	15.6	Km^2
Mean altitude h_m	464.0	m a.s.l.
Main channel slope i	17.5	%
Mean alluvial cone slope S_f	14.0	%
Cohesion c	2500	N/m^2
Angle of internal friction ϕ	31	°
Saturated hydraulic conductivity k_{sat}	10-6	m/s
Volume unit weight γ_s	1430	Kg/m^3

of the rotating ball of the rheometer, cause rapid sedimentation, and large scattering of viscosity readings when the measuring tool is interacting with the particles. To better perform the rheological measurements the samples were sieved with a cut-off of 4 mm (very fine gravel), a method justified by the results of Schatzmann *et al.* [5] and Schatzmann *et al.* [29]. The matrix consists of moderately sorted sand to very fine gravel and low mud content. The grain size distribution of each sample is reported in figure 3. The samples were examined considering a fixed water content representing the actual debris-flow conditions and in fully water-saturated conditions, tab. 2.

Figure 2: Manfredonia after the July 1972 storm event (a); map of the inundated areas (Bissanti [31]) (b).

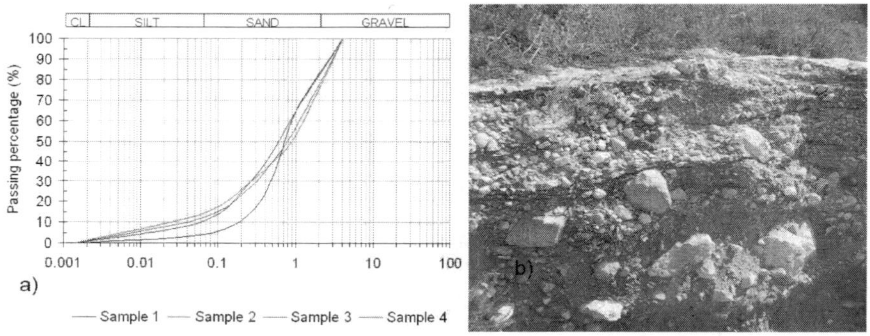

Figure 3: Grain size distribution of the sample matrices collected in the Pulsano watershed (a); section of debris-flow deposits (b).

Table 2: Sediment concentrations of the samples.

Sample	Location	Sediment concentration by volume C_v	
1	Depositional area	0.42	0.78
2	Depositional area	0.42	0.80
3	Source area	0.42	0.78
4	Source area	0.42	0.73

In order to calculate the total debris mobilizable volume a 10 m grid DEM of the Pulsano watershed was used. Using field surveys and photo interpretation, the thickness of the deposits prone to landsliding was evaluated and combined with the results of the stability model SHALSTAB to get the potentially unstable volumes. The inflow debris-flow hydrograph was computed by means of a rainfall-runoff analysis and combined with a variable distribution of the sediment concentration. The computational domain used for the 2D debris-flow modeling was obtained from a 90 m grid DEM. A second 5 m grid system was used to define the cross-section geometry of the torrent and a road network was added to complete the map.

3 Results

The rheological parameters (viscosity η and shear stress τ) of the samples collected in the Pulsano watershed were estimated for different sediment concentrations C_V. At C_V=0.42 eight different shear rates (0.1, 0.3, 0.5, 1, 3, 5, 10, and 30 s^{-1}) were performed. The samples showed a similar shear-thinning flow behavior (viscosity decreases with shear rate). Plotting the average curve of all samples, a good estimate of the viscosity of the debris-flow material was obtained. In water-saturated conditions ($C_V \cong 0.8$) four different shear rates (0.1, 1, 10, and 30 s^{-1}) were performed. The results confirmed the shear-thinning flow behavior of the mixtures.

The Bingham model was fitted to the shear stress data to derive the Bingham viscosity and yield stress. At shear rates lower than 0.1 s^{-1} measurements were influenced by sedimentation (the tool is then measuring in the more dense lower part of the sample) so they were excluded from the dataset. Figure 4 reports the viscosity (a) and shear stress (b) experimental values, obtained for C_V = 0.42 and C_V = 0.8, varying the shear rate. The figure also contains the Bingham viscosity and yield stress calculated for both the sediment concentrations.

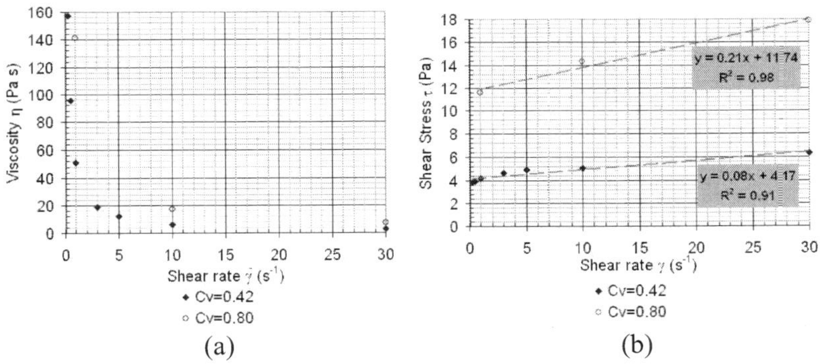

Figure 4: (a) Viscosity and (b) shear stress versus shear rate for tests with different C_V. The Bingham model fitted to the data.

The empirical relationships between viscosity *η* and C_V and between yield stress τ_y and C_V were compared with those found in literature, fig. 5. At sediment concentrations C_V ÷ 0.3-0.4 viscosity and yield stress show different values on the basis of the sediment composition. For example, O'Brien and Julien's matrices [32], contain only fine sediments (d < 63 μm) and samples having lower clay contents are characterized by lower viscosity and yield stress at the same concentration. Rheological parameters obtained by Hübl and Steinwendtner [33] are related to viscous debris-flows with maximum grain size up to 20 mm. And, finally, the material investigated by Kang and Zhang [34] is a poorly-sorted silt for which a lower sensitivity of the rheological parameters to the water content was observed. In this study the viscosity and yield stress values are generally lower, at the same total sediment concentration C_V, than those found in literature and present a low sensitivity to the water content. However, these results are in agreement to the others as the samples have high percentages of sand and low percentages of finer fractions.

Using the mean experimental values of the rheological parameters and the FLO-2D model the inundated areas and water-sediment depths of the 15 July 1972 event were computed. The debris-flow reconstruction is consistent with the estimations based on the existing documentation, fig. 6. The maximum flow depths simulated by the model are quite similar to those reported by the historical map, i.e. h > 1.25 m.

The rheological parameters presented in figure 5 (Bisantino *et al.* [6]) have been used in debris-flow modeling to represent the real inundated area. The results justify the grain size cut-off in rheological measurements and the use of the BMS device when debris matrix contains high percentages of sandy particles.

Further modeling was carried out in order to evaluate the depositional features of the coarser fraction. Generally sediment rheology is heterogeneous from head to tail: coarse clasts are concentrated at the head, are relatively dry

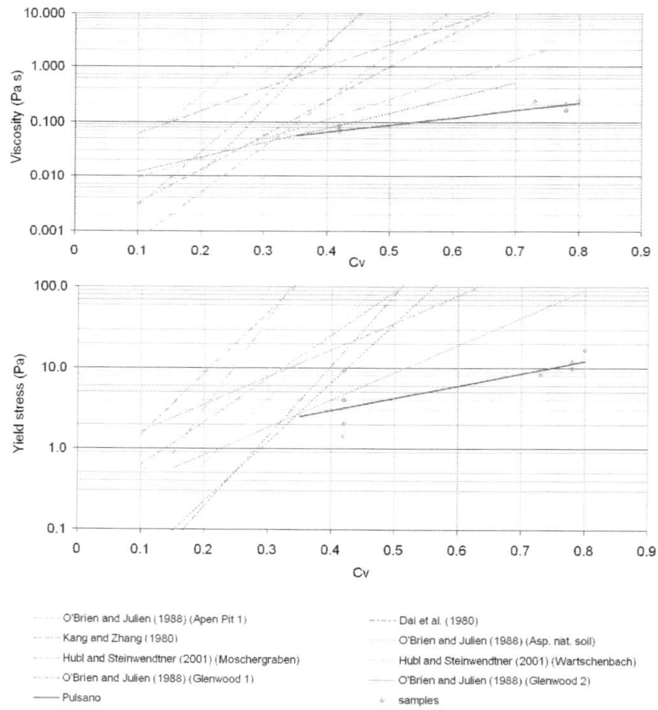

Figure 5: Viscosity and yield stress dependency on total sediment concentration C_V.

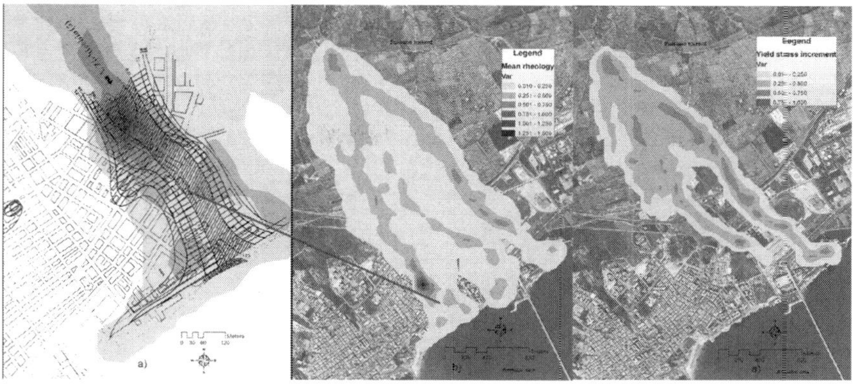

Figure 6: Debris-flow reconstruction of the 15 July 1972 event: (a) comparison between observed and simulated inundated area; (b) overall calculation. The calculations are based on the rheological parameters obtained with the BMS (Bisantino *et al.* [6]); (c) debris-flow simulations with increasing yield stress.

and provide much frictional resistance; a more liquefied material follows the head and exhibits lower yield strength (Iverson [35]). The coarse-grained fraction is mainly dominated by frictional grain interactions before the deposition (Sohn [36]). In the modeling the relationship between yield stress and sediment concentration was modified according to O'Brien and Julien [32], (Aspen Pit 1). This assumption determines the increment of yield stress from $\tau_y \cong 2.5\ Pa$ to $\tau_y \cong 100\ Pa$. The consequence is that the inundated areas are lower and the deposits are concentrated near the stream network, fig. 6, as observed for the coarser material during the survey carried out after the 15 July 1972 event.

4 Conclusions

In this study the dynamics and the rheological properties of the debris-flows in the Pulsano watershed (Southern-Italy) were investigated. The estimated debris-flow hydrograph that reproduce the July 1972 event was derived from rainfall-runoff analysis and the application of the stability model SHALSTAB.

The debris-flow material has a sand content up to 80% and lower fractions of gravel and fine sediments. A rheometrical tool (BMS ball measuring system) for large particle suspensions up to 4 mm was chosen to analyze the matrices (Bisantino *et al.* [6]). The measurements were performed, at different sediment concentrations, on some samples taken from the source and depositional areas. The experimental viscosity and yield stress have a lower sensitivity to the sediment concentrations as the samples contain high amount of particles having grain size d > 63 μm.

The FLO-2D code used to simulate the debris-flow event well estimated the flow depths and the inundated areas when using the rheological data obtained with the BMS. This result confirms the importance of considering the effects of the sandy fraction that mainly constitutes the debris matrix. Yield stress was then increased to evaluate the behavior of the coarse fraction. Depositional areas, simulated by the model, were consistent to those observed after the 15 July 1972 event. The approach adopted in this study can be useful as a part of the analysis involved in the design of the mitigation structures.

References

[1] Gentile F., Bisantino T. & Trisorio Liuzzi G., *Debris-flow risk analysis in south Gargano watersheds (Southern Italy)*. Natural Hazards 44(1), 2008, 1-17.

[2] Montgomery D.R. & Dietrich W.E., *A physically based model for topographic control on shallow landsliding*. Water Resources Research 30(4), 1994, 1153-1171.

[3] O'Brien J.S., Julien P.Y. & Fullerton W.T., *Two-Dimensional Water Flood and Mudflow Simulation*. Journal of Hydraulic Engineering 119(2), 1993, 244-261.

[4] Gentile F., Bisantino T., Puglisi S. & Trisorio Liuzzi G., *Analysis and modeling of debris-flows in Gargano watersheds (Puglia Region - Southern Italy)*. In Lorenzini G, Brebbia CA, Emmanouloudis DE (eds): Monitoring, Simulation, Prevention and Remediation of Dense and Debris-flows. Series: WIT Transactions on Ecology and the Environment, Wit press 90, 2006, 181-191.

[5] Schatzmann M., Fischer P. & Bezzola G.R., *Rheological behaviour of fine and large particle suspensions*. Journal of Hydraulic Engineering 129, 2003, 796-803.

[6] Bisantino T., Fischer P. & Gentile F., *Rheological characteristics of debris-flow material in South- Gargano watersheds*. Nat Hazards, 2009, DOI 10.1007/s11069-009-9462-4.

[7] Takei., A., *Interdependence of sediment budget between individual torrents and a riversystem*. In: Proceedings of the international symposium Interpraevent, Villach, Austria, 5-8 June 1984.

[8] Kronfellner-Kraus., G., *Quantitative estimation of torrent erosion*. In: Proceedings of the international symposium on Erosion, Debris-flow and Disaster Prevention, Tsukuba, Japan, 3-5 September 1985.

[9] Public Works Research Institute., *Technical standards for erosion and sediment control: Basics of planning measures against debris-flow. Planning countermeasure facilities against debris-flow*. Ministry of Construction of Japan, Kyoto, 1984.

[10] D'Agostino V., Cerato M. & Coali R., *Il trasporto solido di eventi estremi nei torrenti del Trentino Orientale*. In: Proceedings of the international symposium Interpraevent, Garmisch-Partenkirchen, Deutschland, 24-28 Giugno 1996.

[11] Marchi L. & Tecca P. R., *Magnitudo delle colate detritiche nelle Alpi Orientali Italiane*. Geoingegneria Ambientale e Mineraria 33(2/3), 1996, 79-86.

[12] Marchi L. & D'Agostino V., *Estimation of debris-flow magnitude in the eastern Italian Alps*. Earth Surface Processes and Landforms 29, 2004, 207-220.

[13] Tropeano D. & Turconi L., *Valutazione del potenziale detritico in piccoli bacini delle Alpi Occidentali e Centrali*. CNR-IRPI/GNDCI Pubblicazione n 2058, Torino, Italy, 1999.

[14] Hungr O., Morgan G.C. & Kellerhals R., *Quantitative analysis of debris torrent hazards for design of remedial measures*, Canadian Geotechnical Journal 21, 1984, 663–677.

[15] Pack R. T., Tarboton D. G. & Goodwin C. N., *GIS-based landslide susceptibility mapping with SINMAP*. In: Bay J. A. (eds) Proceedings of the 34th symposium on Engineering Geology and Geotechnical Engineering, Logan, Utah, 1999.

[16] Hammond C., Hall D., Miller S. & Swetik P., *Level I stability analysis (LISA)*. Documentation for Version 2, General Technical Report INT-285, USDA Forest Service Intermountain Research Station, 1992.

[17] Iverson R. M., *Landslide triggering by rain infiltration*. Water Resources Research 36(7), 2000, 1897-1910.

[18] Rickenmann D., *Empirical relationships for debris-flows*. Natural Hazards 19(1), 1999, 47-77.

[19] Brochot S., Marchi L. & Lang M., *Debris-flow volume assessment: available methods and application to the Poucet torrent (Savoy, France)*. Bulletin of Engineering Geology and the Environment 61(4), 2002, 389-402.

[20] Arattano M., Franzi L. & Marchi L., *Influence of rheology on debris flow simulation*. Natural Hazards and Earth System Sciences, 6, 2006, 519–528.

[21] Sosio R., Crosta G.B., & Frattini P., *Field observations, rheological testing and numerical modeling of a debris-flow event*. Earth Surface Processes and Landforms, 32, 2007, 290–306.

[22] Coussot P. & Piau J.M., *A large-scale field concentric cylinder rheometer for the study of the rheology of natural coarse suspensions*. J. Rheol. 39, 1995, 105-124.

[23] Malet J.P., Remaître A., Maquaire O., Ancey C. & Locat J., *Flow susceptibility of heterogeneous marly formations: implications for torrent hazard control in the Barcellonette Basin (Alpes-de-Hhaute-Provence, France)*. In: Rickenmann and Chen (eds): Debris-flow Hazards Mitigation: Mechanics, prediction and Assessment. Milpress, Rotterdam, 2003, 351-362.

[24] Locat J., *Normalized rheological behaviour of fine muds and their flow properties in a pseudoplastic regime*. In: Cheng-lung Chen (Ed.), Debris-Flow Hazard Mitigation: Mechanics, Prediction, and Assessment, Proceedings of First International Conference, Water Resources Division, American Society of Civil Engineers, 1997, 260–269.

[25] Bagnold R.A., *Experiments on a gravity-free dispersion of large solid spheres in a Newtonian fluid under shear.* Proceedings, Royal Society of London, Series A, Vol. 225, 1954, 49-63.
[26] Coussot P., Laigle D., Arattano M., Deganutti A. & Marchi L., *Direct determination of rheological characteristics of debris-flow.* Journal of Hydraulic Engineering, 124(8), 1998, 865-868.
[27] Hübl J. & Steinwendtner H., *Estimation of rheological properties of viscous debris-flow using a belt conveyor.* Phys. Chem. Earth (B) 25(9), 2000, 751-755.
[28] Wang J.S. & Jan C.D., *Rheological Behavior of Gravel-Mud Mixtures.* 5th International Conference on Hydro-Science & Engineering Warsaw, Poland, 2002.
[29] Schatzmann M., Bezzola G.R., Minor H.E., Fischer P. & Windhab E.J., *Rheometry for large particulated fluids: 2 Comparison of rheometry for debris-flow materials.* Rheologica Acta, 48(7), 2009, 715-733.
[30] Kaitna R., Rickenmann D. & Schatzmann M., *Experimental study on rheologic behaviour of debris-flow material.* Acta Geotechnica 2, 2007, 71-85.
[31] Bissanti A. A., *L'alluvione del luglio 1972 a Manfredonia.* In: Collana Memorie dell'Istituto di Geografia-Facoltà di Economia e Commercio, vol 5. Università degli Studi di Bari, Italy, 1972.
[32] O'Brien J.S. & Julien P.Y., *Laboratory analysis of debris-flow properties.* Journal of Hydraulic Engineering, 114 (8), 1988, 877-887.
[33] Hübl J. & Steinwendtner H., *Two-Dimensional simulation of two viscous debris-flows in Austria.* Phys. Chem. Earth (C) 265(9), 2001, 639-644.
[34] Kang Z. & Zhang S., *A preliminary analysis of the characteristics of debris flow. Proceedings of the International Symposium on River Sedimentation.* Chinese Society for Hydraulic Engineering: Beijing. 1980, 225-226.
[35] Iverson R. M., *The physics of debris flows.* Rev. Geophys., 35, 1997, 245-296.
[36] Sohn Y.K., *Coarse-grained debris-flow deposits in the Miocene fan deltas, SE Korea: a scaling analysis.* Sedimentary Geology, 130, 2000, 45-64.

Formation, expansion and restoration of a sedimentation fan: the case of the Arroyo del Partido stream (Spain)

J. A. Mintegui Aguirre[1], J. C. Robredo Sánchez[1], L. Mao[2] & M. A. Lenzi[3]
[1]*ETSI Montes, Universidad Politécnica de Madrid, Spain*
[2]*Department of Geography, University of Hull, UK*
[3]*Dipartimento Territorio e Sistemi Agro-Forestali, Università di Padova, Italy*

Abstract

The Arroyo del Partido is a small stream flowing into the Doñana National Park marsh. In its lower part, the stream used to flow in a wide, low-gradient area, where the channel was highly dynamic with frequent avulsions. As a result of a channel rectification, the subsequent floods heavily eroded the artificial banks and created over time a depositional fan within the Doñana marsh. The bedload dynamic at the annual scale was estimated from the expansion of the sedimentation fan, assessed from aerial photographs analysis and repetitive topographic surveys. The bedload transport ranged from 3.7×10^6 to 2.1×10^3 m^3 per year, depending on the sequences of occurring floods and on the changing geometrical conditions of the main channel. The progressive erosion of the artificial banks led to self-established cross-sections and slope geometry, which reduced shear stress acting on the bed and favoured sediment deposition on the floodplains, reducing overtime the expansion of the sedimentation fan. This evidence has been used to plan a restoration strategy for the fan and floodplains, aimed to mitigate the effect of the embankment and to improve the overall functionality of the stream. As a result of the restoration, an artificial levee was eliminated and a wide floodplain area was recreated as a free-meandering floodplain.

Keywords: channel rectification, sedimentation fan, sediment transport, aerial photographs, topographical surveys, channel restoration.

WIT Transactions on Engineering Sciences, Vol 67, © 2010 WIT Press
www.witpress.com, ISSN 1743-3533 (on-line)
doi:10.2495/DEB100211

1 Introduction

The morphology and the dynamics of an alluvial river channel are the consequence of transport/deposition of sediments in the river bed. The component of the sediments transported mostly in contact with the bed is called bedload, and is the portion of the total sediment load that determines channel morphology. An accurate estimation of sediment transport is thus important for the prediction of channel morphological changes as well as to the success of a variety of river engineering and management practices such as channel design, river regulation, and in-stream construction. Bedload transport formulas are usually calibrated on laboratory data or on a limited amount of observations on specific study rivers, reducing the chances of their reliable application to a certain field site. Bedload formulas often overestimate sediment transport rates by one or more orders of magnitude, and sediment rates are accurately predicted only under unlimited sediment availability and supply conditions [4]. On the other hand, direct measurements of bedload transport requires a considerable field effort and are difficult to be conducted especially during floods, even if recent efforts have been made in developing and testing advanced field methods for the direct and indirect quantification of bedload transport [e.g. 1, 10, 19]. Additionally, representative measurements are difficult to obtain due to the significant spatial and temporal variability associated with bedload movement [6]. As an alternative, the rate of bedload transport can be evaluated at the single flood or at longer time scales using a morphological approach, i.e. measuring erosion or deposition volumes over a certain period [7, 9, 12, 13]. This approach can be also used to test bedload formulas over long temporal and large spatial scales [2, 12, 17]. Recently, Pelpola and Hickin [18] estimated the volumetric expansion of a small delta into a lake using sequential aerial photography, bathymetry, and ground-penetrating radar surveys over a period of 52-year, confirming the efficiency of the morphological approach for determining long-term bed load transport rates.

The aim of the present work is to present the results of a morphologically based quantification of bedload transport in a small Spanish stream (Arroyo del Partido) where a channel rectification caused the formation of a sedimentation fan into a protected marsh. The paper also presents the restoration project carried out along the most impacted reach of the stream in order to reduce the fan growing and to improve the overall functionality of the stream.

2 Study site and its recent record of human impacts

The Arroyo del Partido is a low gradient basin located in Southwest Spain (Andalucia region, Huelva province). For its last 6 km, the stream crosses the Natural Park of Doñana (UNESCO heritage site since 1994), finally discharging into Doñana marsh inside the National Park (Fig. 1). The maximum and minimum heights of the basin are 121 and 5 m a.s.l. respectively, and the surface area is of 308 Km^2. In its lower portion, the slope of the channel does not

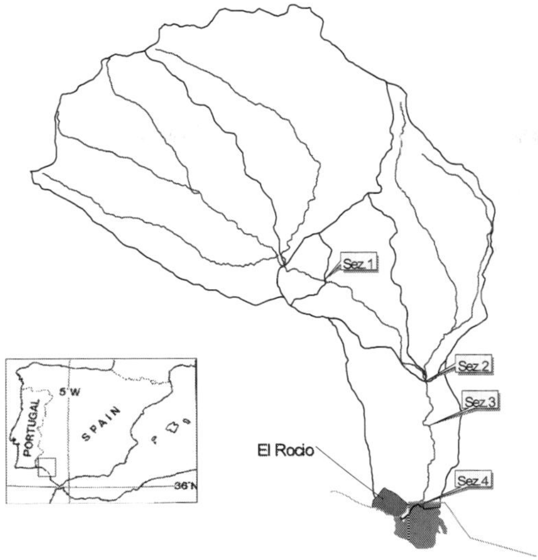

Figure 1: Location and map of the Arroyo del Partido basin and sub-basins. In 1981 the stream was chanalized from section 2 to section 4.

exceed 1% [14, 15], and the channel bed and floodplains are essentially characterized by a fine and nearly homogeneous sand (D_{50} = 3 mm; D_{84} = 6 mm).

Even if the Doñana wetland is connected to the Atlantic Ocean, the Arroyo del Partido basin is characterized by a Mediterranean-type climate. The average rainfall is about 660mm per year and 80% of precipitations occur from October to March. Intense rainfalls up to 60mm per day are common and characterize the torrential regime of the Arroyo del Partido stream.

Until 1980, the lower portion of the Arroyo del Partido used to flood in a wide floodplain area (up to 1 km) before reaching the Doñana marsh (Fig. 2). Within this floodplain area (3700 ha) the stream was free to create avulsions and meanders as well as depositing sediments. Abandoned channels in the floodplain suggest that the original Arroyo del Partido had irregular cross section up to 10 m wide, 2 m deep and had a slope of about 0,165%.

In 1981 the lower 8 km long reach was rectified in order to cultivate the original sandy floodplain (Fig. 2), which was levelled and converted into arable land. The channel was that narrow (15 m) that early '80 flood events caused a severe incision of the channel and removed most of the embankments. In 1985 a wider channel (50 m) was built. Following floods (especially the 1995-1998 winter events) seriously affected the channel embankments because of erosion of both stream banks and bed. As a consequence of the increased sediment transport, a sedimentation fan was produced at the closing section (see Results), which correspond to the Doñana marsh very near the El Rocio village. An attempt of stabilizing the bed by building a regular meandering channel within a 100m wide floodplain (Fig. 2) unsurprisingly failed because of repeated massive sedimentation of the whole structure.

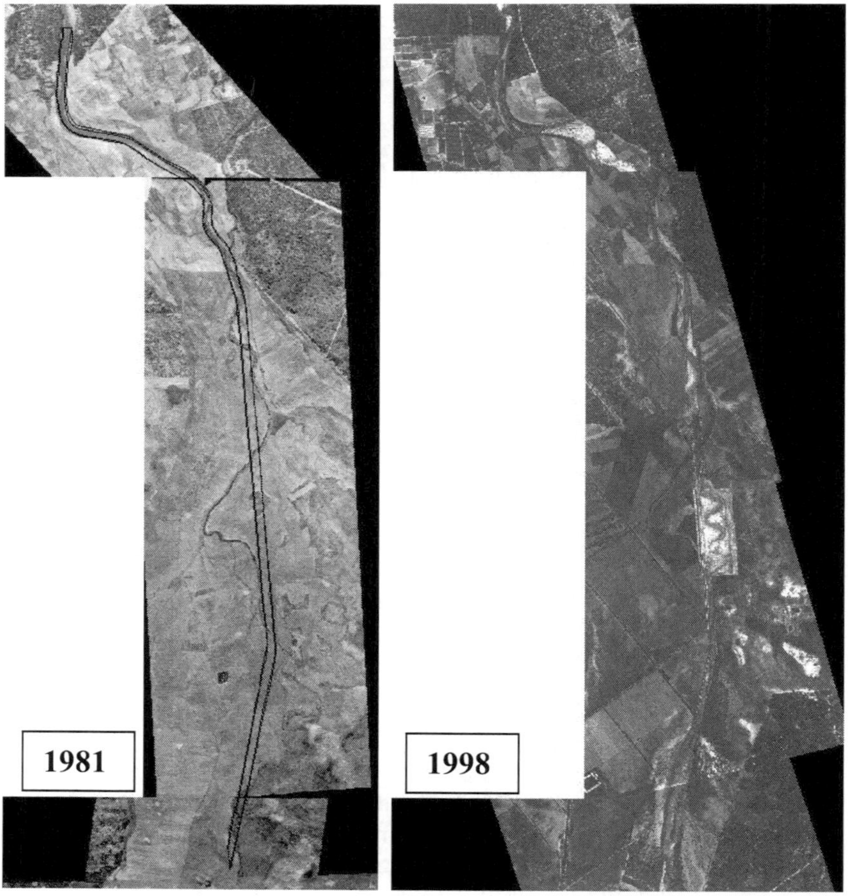

Figure 2: The lower reach of the Arroyo del Partido in its quasi-natural conditions as appeared in an aerial photo from 1956 (photo on the left). The black lines define the borders of the channel built in 1981. The rectification caused severe bank erosion, which resulted in overflows and abundant sediment deposition on the floodplain (photo on the right).

3 Methods

The planform expansion of the sedimentation fan was determined by comparing aerial photographs of the fan taken at certain intervals. Eleven aerial photographs of the lower portion of the Arroyo del Partido stream and of its fan (Fig. 3) were obtained or specifically committed (1954, 1982, 1985, 1993, 1996, 1997, 1998, 2000, 2001, 2001 and 2003).

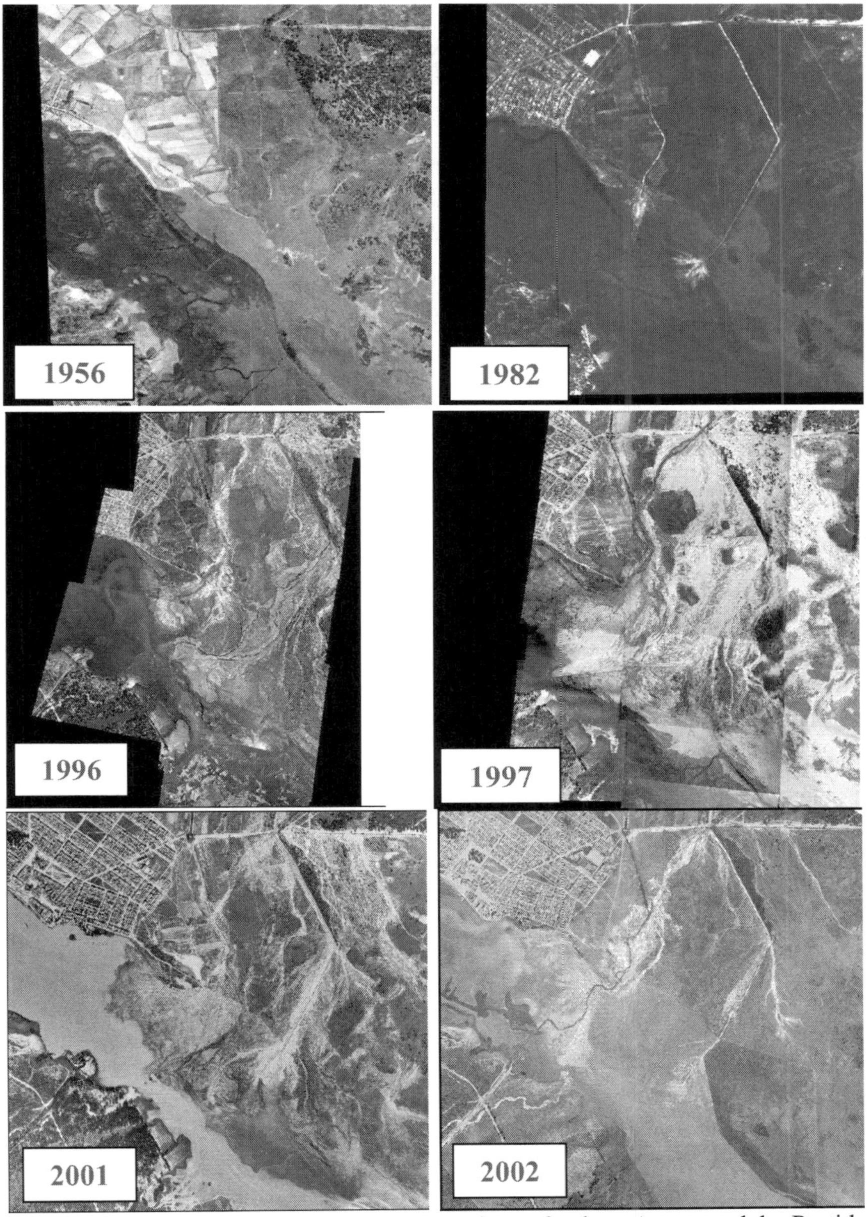

Figure 3: Successive aerial photographs of the Arroyo del Partido sedimentation fan within the Doñana marsh. The lower portion of the channel rectification is clearly visible in the 1982 photo, as well as the incipient accumulation of transported sand at its downstream end. The sedimentation fan experienced a formidable planform increase in the late 1990s.

The aerial photographic prints have been scanned at a resolution of 300 to 600 dpi in order to obtain an average pixel dimension of less than 2 m depending on the scale of the photos. The aerial photographs have then been geocorrected and co-registered at a common mapping base at 1:10000 scale. A minimum of 10 clearly visible control points have been used to warp single images. Second order polynomial transformations have been then applied, obtaining root mean square errors (RMSE) always less than 5 m. The margin of the sedimentation fan was digitized on each aerial photograph (using Arc View v. 3.2) and was identified due to the distinct colour changes between the sediment of the fan and the vegetation of the marsh. The aerial photographs were always taken in summertime.

A discharge gauging station built on the middle portion of the canalized reach of the Arroyo del Partido have continuously suffered from sedimentations and provided just few isolated values. This have been used to calibrate the application of an hydrological model (HEC-HMS v.2.1.3.) to generate continuous discharge from continuous hourly precipitation records collected in a raingauge station located within the basin [11, 15].

4 Expansion of the sedimentation fan over the Doñana marsh

The margins of the sedimentation fan were digitized on aerial photographs and were used to calculate the fan areas and expansion between two successive analyzed years (Fig. 4). As to the quantification of the volumetric increase of the fan into the Doñana marsh, three topographical surveys were carried out in summer 1997, 1998 and 2003. A total station was used on the first two surveys, whereas a differential GPS device was used in the later. The acquired topographic data allowed the calculation of digital elevation models of the whole area for the mentioned three years. To reconstruct the digital elevation model of the area before the canalization, a series of core samplings were taken during the 1997 topographical survey and allowed the estimation of the depth of deposited sand over the clayey marsh surface. This allowed a reasonable estimation of the volume of the sediments deposited over the Doñana marsh in 1997 ($\approx 2.5\times10^6$ m^3), 1998 ($\approx 3.7\times10^6\,m^3$), and 2003 ($\approx 4.4\times10^6\,m^3$).

The volumetric annual expansion of the fan was then estimated assuming a linear relation between the aerial and the volumetric expansion of the sedimentation cone. The results show that the fan virtually appeared only in 1993, thus more than a decade after the channel rectification (Fig. 5). Even if in this period a series of high magnitude floods occurred, these affected the upper canalized portion of the stream (channel erosion, banks removal, and sedimentation on the floodplains) without causing substantial amount of sediments reaching the Doñana marsh. In fact, during this first phase (1981-1993) the annual transport rate was very low ($2.1\times10^3\,m^3$ per year). Soon after that, the sedimentation cone experienced a dramatic expansion and reached a volume of a $3.7\times10^6\,m^3$ by 1998, at a rate of $7.4\times10^5\,m^3$ per year from 1993 to 1998 (Fig. 5). The highest transport rate was registered between 1996 and 1997,

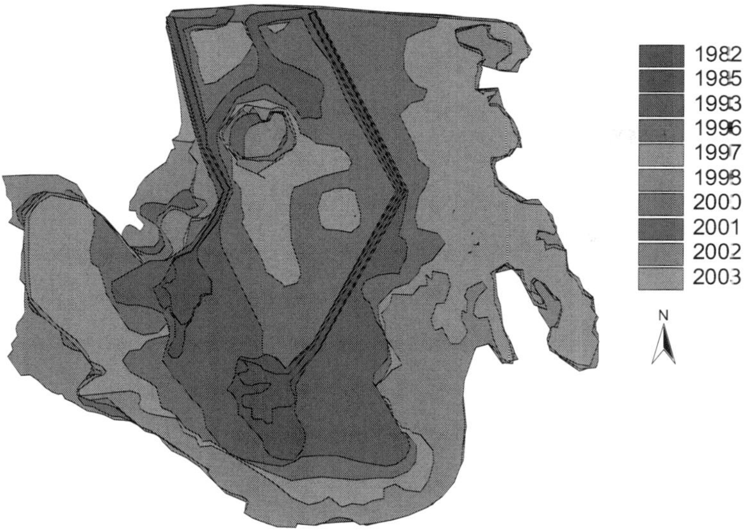

Figure 4: Planimetric expansion of the Arroyo del Partido fan on the Doñana marsh from 1982 to 2003. The upper border corresponds to a road path with fords reaching the El Rocio Village.

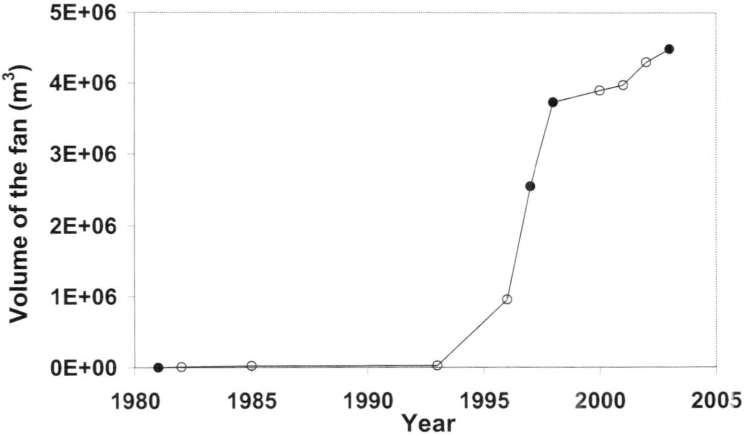

Figure 5: Volumetric increase of the Arroyo del Partido fan. The solid circles represent actual volumes measured in the field, whereas the empty ones represent the values obtained from the planimetric expansion of the fan.

with a transport rate up to 1.6×10^6 m^3 per year. After 1998 the expansion rate decreased substantially stabilizing around 1.5×10^5 m^3 per year from 1998 to 2003.

The variability associated with average annual bedload rate is partially due to the occurrence and nature of flood events. In fact, 1996, 1997, and 1998 (when the fan experienced the fastest volumetric expansion) where characterized by the higher number of bedload transport events, and by the occurrence of the higher magnitude flood (277 m^3 s^{-1}). However, the following decrease of bedload transport can be also ascribed to morphological changes experienced by the channel bed which reduced its transport capacity. The average channel slope, which was around 0.165% before the canalization, was still as steep as 0.2% during 1999 and reduced to 0.157% in 2003. This is likely due to the progressive expansion of the sedimentation fan and to an increased tendency of meandering. Moreover, by 1998 most of the artificial levees were eroded, and the over-bankfull flows were able to flood and deposit sand on the adjacent floodplains. Considering the shape of cross-sections measured from 1981 to 2003, it has been estimated that, relative to the pre-canalization conditions, the bankfull discharge (recurrence interval 1.2 years) flowed with a shear stress six times higher just after the canalization. Due to later self-established cross section geometry with increased channel width, by 2003 the bankfull shear stress would have been lower, i.e. four times higher than in the pre-1981 conditions [11].

Overall, the combined effects of the reduced shear stress in the channel and the progressive stabilization of channel bars and banks by vegetation has progressively reduced the sediment availability, thus decreasing sediment transport from 1993 to 2003.

5 Restoration of the sedimentation fan

The assessment of the annual bedload transport amount and dynamics from 1981 to 2003 was functional to the definition of a management and restoration strategy for the Arroyo del Partido sedimentation fan. This need was driven by the fact that the Doñana National Park (classified as RAMSAR and UNESCO world heritage site) represents the main protected area in Spain. The overall objective of the restoration of the Arroyo del Partido was to detain its fan from growing more deeply into the Doñana marsh and to recover abandoned farming lands to provide menaced Imperial eagle and Iberian lynx populations with a suitable hunting ground [20].

In order to detain the growing fan and to restore the portion of canalized stream, Mintegui et al. [14] proposed to remove one of the banks and to return an area of 1500 ha to the stream as a natural floodplain. The aim was to facilitate the undergoing tendency of the main channel to widen and reduce its slope and shear stress during floods by favouring overbank flowing and sand deposition on the floodplain. This would have led to a high degree of freedom for the stream to wander, meander and create avulsions within the floodplain, as well as stopping the fan expanding into the Doñana marsh. The restoration plan was implemented in 2006 with the removal of some remaining portions of banks and with the built of a transversal bed sill at the upstream end of the original canalized portion of the stream in order to stabilize the bed and prevent bed incisions. The restoration plan is monitored continuously in both the hydraulic [16] and the vegetation self-reestablishment point of view [3].

6 Conclusions

The paper present an application of the morphological approach for the quantification of bedload transport rate at the annual scale, and demonstrates the utility of combining sequential aerial photography and topographical surveys to reconstruct the planimetric and volumetric expansion of a sedimentation fan. The applied method is relatively inexpensive and easy to use if compared with direct or indirect field devices for the bedload monitoring. Even if the transport rate can be only be assessed at the annual scale, the method allows avoiding errors associated with bedload field survey during high flows and to exclude the influence of short-term fluctuation of bedload [5, 8] in the sediment transport calculation. The method is thus a viable alternative to direct monitoring for assessing the long-term bedload rates in similar conditions worldwide. A similar approach, together with the contemporary use of bathymetry and ground-penetrating radar, has proved to allow the quantification of bedload rates in a partially submerged delta [18].

The annual bedload transport rate ranged from 3.7×10^6 to 2.1×10^3 m^3 per year between 1982 and 2003, and this remarkable variability have been mainly related to the changing geometrical conditions of the main channel. For a few years after the canalization (1981-1993) the stream eroded locally banks and margins, depositing sediments in the overflow areas and carrying little amount of sediments at its downstream end. In a following phase (1993-1998), characterized by a considerable number of high magnitude floods, a substantial amount of sediments reached the Doñana marsh and created a much extended fan (3.8×10^6 m^2). During this period, the sequence of floods progressively favoured the self-establishment of cross-sections and the channel slope adjustment such that over the following years (1998-2003) the fan reduced its expansion rate.

The overall assessment of sediment transport and the evidence that a free wandering channel within a defined floodplain reduced the volumetric expansion of the fan has driven the choice of the restoration strategy for the Arroyo del Partido. The restoration plan, implemented from 2006, entail the removal of an entire artificial bank and make "room" for the river within a protected floodplain area set aside from cropping, where the flow will deposit sediments and possibly create secondary channels.

Acknowledgements

We would like to thank the Scientific Commission of the Project Doñana 2005 and the Conservation Area Manager of the Doñana National Park for their support. The paper has been written while LM was supported by a Marie-Curie fellowship (EU 7th Framework Programme; PIEF-GA-2008-219294).

References

[1] Bunte, K., Abt, S.R., Effect of sampling time on measured gravel bed load transport rates in a coarse-bedded stream. *Water Resources Research*, 41, W11405, doi:10.1029/2004WR003880, 2005.

[2] Carson, M.A., Griffiths, G.A., Bedload transport in gravel-bed channels, *Journal of Hydrology (New Zealand)*, 26(1), pp. 1-15, 1987.

[3] Garcia-Novo, F., Escudero Garcia, J.C., Carotenuto, L., Garcia Sevilla, D., Fernandez Lo Faso, R.P., The restoration of El Partido stream watershed (Doñana Natural Park). A multiscale, interdisciplinary approach. *Ecological Engineering*, 30, pp. 122-130, 2007.

[4] Gomez, B., The potential rate of bed-load transport. *Proceedings of the National Academy of Sciences of the United States of America*, 103(46), pp. 17170-17173, 2006.

[5] Gomez, B., Richard, L.N., Hubbell, D.W., Temporal variations in bedload transport rates associated with the migration of bedforms. *Earth Surface Processes and Landforms*, 14, 135-156, doi:10.1002/esp. 3290140205, 1989.

[6] Gomez, B., Bedload transport. *Earth-Science Reviews*, 31, pp. 89-132, 1991.

[7] Ham, D., Church, M., Bed-material transport estimated from channel morphodynamics: Chilliwack River, British Columbia. *Earth Surface Processes and Landforms*, 25, pp. 1123-1142, 2000.

[8] Hoey, T.B., Temporal variations in bedload transport rates and sediment storage in gravel-bed rivers, *Progress in Physical Geography*, 16, pp. 319-338, doi:10.1177/030913339201600303, 1992.

[9] Lane, S.N., Richards, K.S., Chandler, J.H., Morphological estimation of the time-integrated bed load transport rate. *Water Resources Research*, 31, pp. 761-772, 1995.

[10] Laronne, J.B., Alexandrov, Y., Bergman, N., Cohen, H., Garcia, C., Habersack, H., Powell, D.M., Reid, I., The continuous monitoring of bed load flux in various fluvial environments. In *Erosion and Sediment Transport Measurement in Rivers: Technological and Methodological Advances*, eds. Bogen, J., Fergus, T., Walling, D.E. IAHS Publication 283, International Association of Hydrological Sciences: Wallingford, pp. 134-145, 2003.

[11] Mao, L., Analisi comparativa del trasporto solido di corsi torrentizi in diversi ambiti geografici. Ph.D. Thesis, University of Padova, Italy, 2004.

[12] Martin, Y., Evaluation of bed load transport formulae using field evidence from the Vedder River, British Columbia. *Geomorphology*, 53, pp. 75-95, 2003.

[13] McLean, D.G., Church, M., Sediment transport along lower Fraser River 2. Estimates based on the long-term gravel budget. *Water Resources Research*, 35, 2549-2559, 1999.

[14] Mintegui J.A., Robredo J.C., Sendra P.J. *Avenidas torrenciales en el Arroyo del Partido y su incidencia en la marisma del Parque Nacional de*

Doñana. Naturaleza y Parques Nacionales, Serie Técnica, Organismo Autónomo Parques Nacionales, Madrid, pp. 373, 2003.

[15] Mintegui, J.A, Lenzi, M.A., Robredo, J.C, Mao L., *Movilización versus estabilización de los sedimentos en los cursos sometidos a la dinámica torrencial*. Organismo Autónomo Parques Nacionales, Ministerio de Medio Ambiente, Serie Técnica, Madrid, pp. 143, 2006.

[16] Mintegui, J.A, Robredo, J.C, De Gonzalo Aranoa, C., Huelin Rueda, P. Restoration of a degraded torrential stream by means of flood control system: the case of Arroyo del Partido stream (Spain). Proceedings Debris Flow conference 2010, this issue.

[17] Nicholas, A., Modelling bedload yield in braided gravel rivers. *Geomorphology*, 36, 89-106, 2000.

[18] Pelpola, C.P., Hickin, E.J., Long-term bed load transport rate based on aerial-photo and ground penetrating radar surveys of fan-delta growth, Coast Mountains, British Columbia. Geomorphology, 57(3-4) pp. 169-181, 2004.

[19] Rickenmann, D., McArdell, B.W., Continuous measurement of sediment transport in the Erlenbach stream using piezoelectric bedload impact sensors. *Earth Surface Processes and Landforms*, 32, pp. 1362-1378, 2007.

[20] Saura Martinez, J., Bayan Jardin, B., Casas Grandes, J., Ruiz deLarramendi, A., Urdiales Alonso, C., Documento Marco para el Desarrollo del Proyecto Doñana 2005. Ministerio de Medio Ambiente, Madrid, pp. 201, 2001.

Dynamics of changes of bed load outflow from a small glacial catchment (West Spitsbergen)

W. Kociuba, G. Janicki & K. Siwek
Institute of Earth Sciences,
Maria Curie-Skłodowska University in Lublin, Poland

Abstract

Fluvial transport in a small glacial river catchment localized in the NW part of Wedel-Jarlsberg Land (Spitsbergen SW) was studied in the summer period of 2009. The intensity of bed load transport was determined using the River Debris Traps constructed for the project's need. The obtained results indicate high dynamics of bed load transport, the amount of which reached up to several dozen kg for 24 hours in individual measurement sites. The results also confirmed great variability of different fluvial processes in the polar zone.
Keywords: small glacial river catchment, bed load transport, bed load sampling, River Debris Trap, Spitsbergen.

1 Introduction

The former investigations of fluvial transport in the polar catchment of NW Spitsbergen included analysis of solution and suspension transport, without the measurements of bed load material's transport (Rachlewicz [9], Chmiel et al. [4]) – the main component of transport in a gravel-bed river (Hammer and Smith [6], Warburton [12]). The mechanism and intensity of bed load transport and functioning of proglacial rivers are poorly investigated in this environment. It is mostly the result of a shortage of representative and long-term measuring series as well as standardisation of measurement methodology. The use of direct methods to measure the bed load transport, just as in mountain gravel-bed rivers, is limited by heavy terrain conditions (Bunte et al. [3]). The use of the advanced measurement techniques (luminescence, magnetic, radio-emitters and radioisotope methods) is not possible due to a shortage of infrastructure, particularly energetic (Froehlich [5]). Therefore, elaboration and the putting into

WIT Transactions on Engineering Sciences, Vol 67, © 2010 WIT Press
www.witpress.com, ISSN 1743-3533 (on-line)
doi:10.2495/DEB100221

practice of the new methodology of bed load transport measurement, as well as determination of the elementary value of bed load concentration, were very important research tasks.

The following study is a preliminary elaboration of the investigations concerning the dynamics of changes of bed load transport in a sub-polar gravel-bed river, which started in 2009. The studies of fluvial processes were conducted in the summer season (July-September) of 2009 during the 21st Polar Expedition of the Maria Curie-Skłodowska University to Spitsbergen. The investigations were supported by grant from the vice-chancellor for Scientific Research and International Cooperation of the Maria Curie-Skłodowska University.

2 Study area

For detailed analysis we chose the catchment of the Scott River (Scottelva), situated in the proximity of the "Calypsobyen" research station of the Maria Curie-Skłodowska University (Figure 1). The analysed catchment, located on the Bellsund Bay in the south-western part of the Wedel Jarlsberg Land (SW Spitsbergen), is about 10 km^2 in area. The Scott Glacier, which is at the stage of strong recession, covers more than half of the catchment (57%). This valley glacier is less than 4 km long and from 1.1 to 1.8 km wide. The glacier front retreated at a rate of 6 m/year in the 1960s and 1970s, to a maximum 15 m/year in the 1990s, and about 20 m/year in recent years (2005-2006) (Zagórski et al. [13]).

The Scott River catchment is diverse in respect of geological and geomorphological conditions. The upper part of the catchment is a glaciated mountain valley, surrounded by mountains ridges composed of Proterozoic

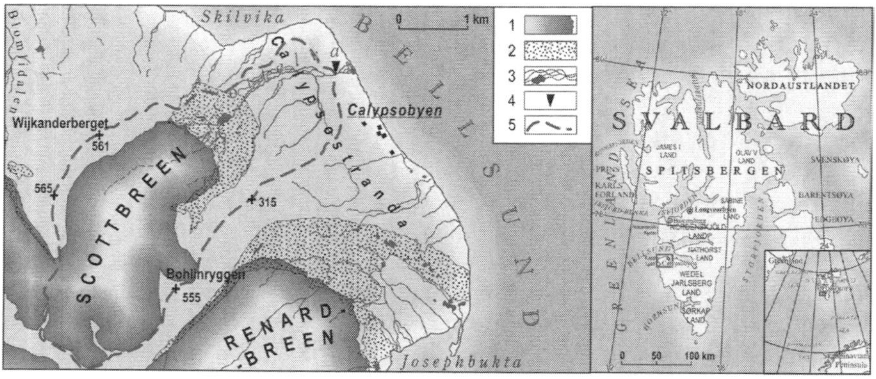

Figure 1: Location of the research area. Situation of the Scott River catchment: 1. valley glaciers, 2. glacial accumulation zones, 3. rivers and water bodies, 4. location of river gauge and research profiles, 5. catchment boundaries.

metamorphic rocks (tillites, phyllites and quartzites). Steep and short rocky slopes are modelled by predominant weathering, nival and mass movement processes. A rather wide zone of glaciomarginal relief occurs near the front of the Scott Glacier. In the extra-marginal zone of the Scott glacier the river flows through the gorge of the frontal moraine and creates a large sandur system. The braided river system with variable channel pattern functions in the central part of the catchment.

The non-glaciated part of the Scottelva catchment includes the shore plain called "Calypsostranda", which was uplifted by isostatic movements at 20-120 m a.s.l. (Szczęsny et al. [11]; Reder and Zagórski [10]. The bedrock of Calypsostranda is composed of Mesozoic deposits (gypsum, dolomite, sandstone, shale and mudstone) covered by Tertiary sandstones with hard coal inserts and Quaternary marine (gravel, sand and clay) and glacial (till) deposits (the thickness of the Quaternary deposits varies from several to 20 m) (Pękala and Repelewska-Pękalowa [8]). In the lower course, the Scott River forms the gorge dissecting the marine terraces (18-40 m a.s.l.). In this section of the valley – several dozen metres long – the braided channels join up into one, and further a system of alluvial fans occurs, which closely adjoin the Recherche fiord. The river mouth is separated from the bay by the storm ridge, and water flows through the crevasse (Figure 1).

The Scott River represents a glacial hydrologic regime and ablation waters dominate in its alimentation structure (90%). The river is also supplied with rainfall (4%), nival (4%) and permafrost (about 2%) waters. The mean annual outflow from the Scott River catchment is about 900 mm, and the mean discharge over a period of many years is estimated at about 900 $dcm^3 \cdot s^{-1}$, which corresponds to the specific discharge of 90 $dcm^3 \cdot s^{-1} \cdot km^2$ (Bartoszewski [1]). The Scottelva is characterized by a great variability of river stages and irregular discharge – daily, seasonal and over a period of many years. The maximum flood discharges caused by the increase in air temperature or/and high precipitation reach 12 $m^3 \cdot s^{-1}$ (7-9.08.1993). The Scott River has several small tributaries supplied with snow-permafrost waters, and the greatest of them is the Renifer Stream draining the eastern part of the sandur (Bartoszewski [1]).

3 Methods of research

The hydrometric station and the traps for fluvial transport measurements were located on the Scott River, in the place where the braided channels join up into one, 350 m from its mouth to the fiord (Figure 1). This location was similar to that of hydrographical monitoring in previous years (Bartoszewski [1], Bartoszewski et al. [2]). The hydrographical research included continuous measurement of the river stage in the channel with the use of a pressure electronic limnograph and analyses of the main physico-chemical features (temperature, conductivity). Measurements of water velocity in the cross profile were made using a hydrometric current meter.

In order to determine the dynamics of bed load transport, the direct measurement methods of dragging intensity were used. The traditional traps are usually operated directly in the river or from a boat, pontoon, footbridge or bridge. The solutions applied at the Polish Hydrometeorological Institute (PIHM), as well as Helley-Smith traps, need the standing presence of a researcher during the measurements. This methodology limits the time of sample collection to short-term periods *ipso facto* results in poorer quality of the obtained results. During high water stages and high suspended sediment concentration there are no possibilities to control a trap arrangement in the channel bed and its lack of stability can disturb the measurement. In the case of PIHM traps, the heavy weight and size of equipment causes limited possibilities to use it in the conditions of high energetic mountain river flow. Bunte et al. [3] partly solve this problem using traps stabilized in the river bed, but it is effective in terms of low and balanced flows only. During the raised water stage servicing these traps requires a minimum of three persons protected with ropes, but changing of the traps is connected with the researchers' immersion in the water. In the mountain rivers environment in the low temperatures this limits the possibilities and time of measurement.

These shortcomings are eliminated by the construction of an all-purpose, light and portable equipment to measure bed load transport in river channels – River Debris Traps [RDT] constructed by W. Kociuba (Figure 2). The advantages of

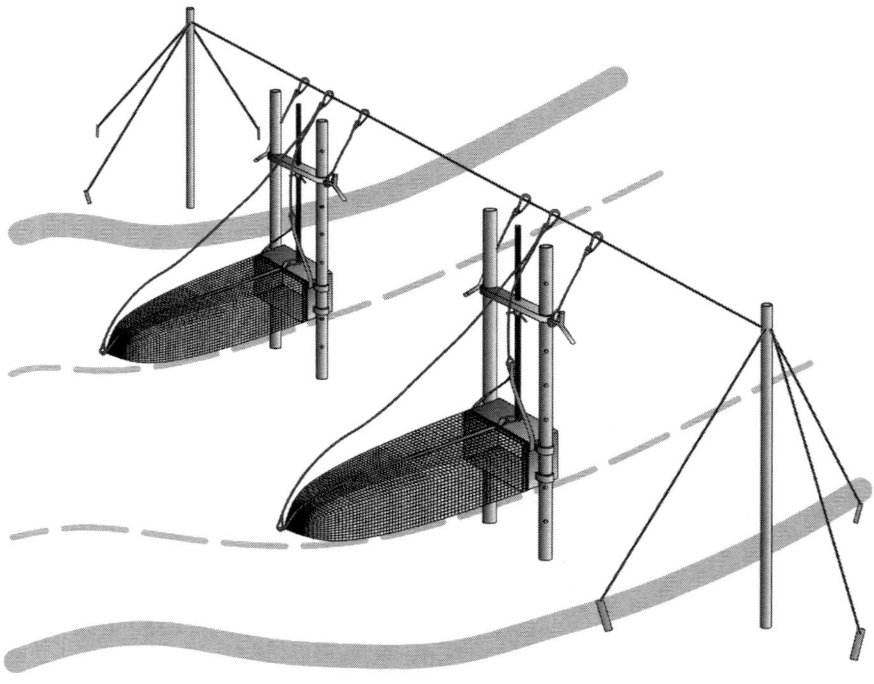

Figure 2: Example of arrangement of two River Debris Trap (RDT) sets and systems of protection in cross section.

this equipment are the following: recurrence of measurements conducted in the time intervals adjusted to the flow regime with integrated monitoring, one-man service and carrying out the measurement independently of environmental conditions in the catchment as well as climate parameters. The use of the set of RDT traps allow one to examine the bed load material movement, which determines the conditions of the commencement of movement and intensity of bed load material transport. The devised measurement method (uninterrupted procedure with samples collected every 24 hours) allows one to determine the interrelations between the discharge rate and elementary bed load transport rate, the grain size distribution of bed load material in the separate measurement profiles and threshold values of movement of individual bed load fractions, as well as main routes of transport.

The working out and implementation of the original method of measurement of bed load transport resulted in the submission of an application for patent protection for the River Debris Trap in December 2009 (no. of application WIPO ST 10/C PL389882).

Sediment transport was measured using the River Debris Traps, which were placed in four measurement sites in the channel cross profile. The caught material was weighed and photographed separately for the each measurement site every 24 hours (Figure 3). River sediment transport was measured from July

Figure 3: The set of river debris traps in the cross profile of the Scottelva. Mass of bed load material caught during 24 hours in one of the measurement sites in the cross profile of the Scottelva (photo by K. Siwek).

10th to September 7th 2009, i.e. from the time when the whole profile of the river channel emerged from snow-ice cornices on the river banks to the time when the channel bed started to freeze. In total 227 samples were taken on 54 measurement days.

The fieldwork also included a geomorphological survey and identification of the permanent and ephemeral macro-, meso- and microforms in the valley and channel. The main parameters of valley floor geometry were determined along the selected research profiles in the selected test areas using a laser tachimeter. The degree of coupling between the slope and channel subsystems was estimated based on the cubature of main (with cross-section > 1 dcm^2) erosion and depositional slope forms.

The main meteorological variables, such as air temperature and precipitation level, were recorded every ten minutes by an automatic meteorological station. The station is situated on the Calypsostranda shore plain (23 m a.s.l.) covered by patch tundra vegetation, about 200 m from the Recherche Fiord shore, near the research station of the Maria Curie-Skłodowska University.

4 Results

In the summer period of 2009 the course of weather and meteorological conditions was such as in other years and no weather anomalies occurred. The air temperature in the measurement period ranged from +2.0°C to +7°C, with the average of +4.9°C (Figure 4). The precipitation total in the measurement period (18.1 mm) was slightly higher than that over a period of many years. During the observed 16 precipitation days, the maximum daily total did not exceed 5 mm

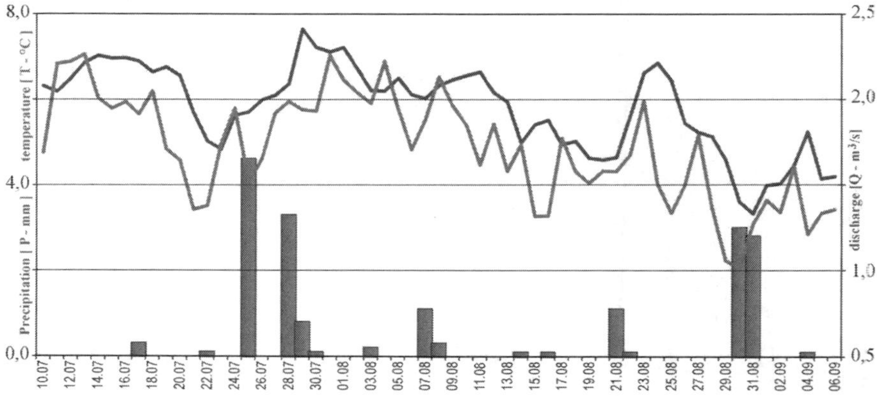

Figure 4: Comparison of the run of changes of average daily air temperature and average daily precipitation total in the "Calypsobyen" research station of the Maria Curie-Skłodowska University with the average daily discharge in the Scott River in the summer season of 2009: H – precipitation, Q – daily discharge, T – average daily air temperature.

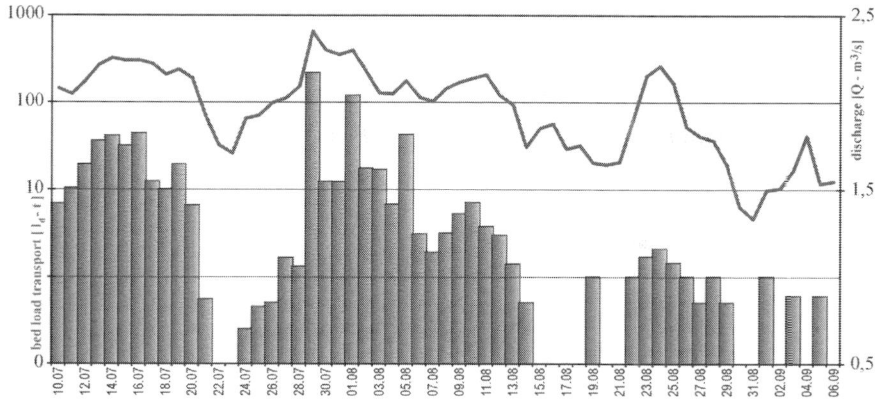

Figure 5: Daily values of bed load concentration (t) and discharge (m^3/s) in the Scott River in the summer season of 2009.

and differed considerably from the previously recorded maxima (36 mm) in 1993 (Bartoszewski et al. [2]).

In the summer season of 2009 the water stage in the Scott River channel ranged from 7 to 22 cm and the average daily discharge (from 1.3 to about 2.4 $m^3 \cdot s^{-1}$) was similar to average values. Both water stages and discharges rather well corresponded to the run of average daily air temperature and daily precipitation total.

The obtained results indicate a great variability of dynamics of bed load transport in the conditions of rather stable discharge: from 1.3 to 2.4 $m^3 \cdot s^{-1}$, with the mean value of 2.0 $m^3 \cdot s^{-1}$, both in daily cycle and in the whole measurement period (Figure 5). The amount of bed load caught in the particular measurement sites ranged from 0 to 66 kg for 24 hours. The mean daily transport in the particular measurement sites was 1.3, 3.8, 3.3 and 4.7, and the average daily value was 3.3 kg. The maximum daily values (from 41.5 to 66 kg) were recorded during the flood on July 29th 2009. These values constituted from 20% to 50% of the total mass measured in the particular measurement sites. The total bed load mass caught in the particular measurement sites was 77, 210, 191 and 275 kg (in total 753 kg). Therefore, the bed load transport in the channel cross profile can be estimated at almost 100 kg for 24 hours, i.e. about 6 t during the measurement period.

The analysis of daily distribution of the bed load values indicates spatial and temporal differentiation of bed load transport (Figure 6). The increase in bed load mass in the particular measurement sites from the southern to the northern river-bank (from S1 to S4) was distinctly marked for the whole measurement period. On particular days the highest daily loads were alternately measured in the sites located in the thalweg zone (S3 and S4), although the daily maximum load (66 kg) was caught in the S2 measurement site. This fact enables us to conclude that transport routes shift in the channel cross profile.

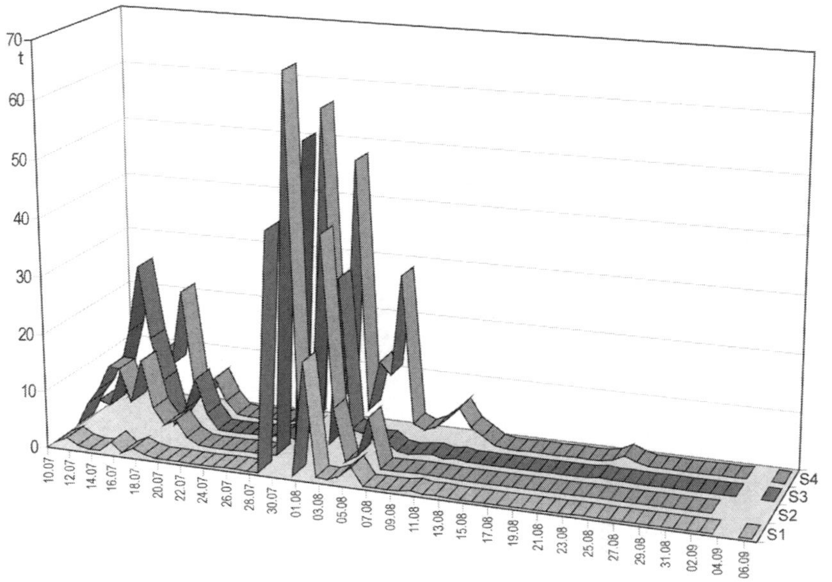

Figure 6: Daily values of bed load concentration (t) in the particular measurement sites of the Scott River cross profile in the summer season of 2009.

In respect of bed load transport variability, the measurement period was distinctly divided into two parts. In the first half (from July 9th to August 9th of 2009) the elementary rate of bed load transport was considerably more variable in the particular sites and days. In the second half (from August 10th to September 7th of 2009) daily bed loads were usually low and not related to discharge values (Figure 6).

The relation between bed load transport rate and discharge changes was different in two parts of the measurement period due to spatial and temporal differentiation of material supply to the channel. In the first part of the measurement period the channel was supplied with slope material redeposited directly to the channel. With the melting of snow patches the mass movement processes were triggered on the southern side of the valley gorge section. This phenomenon resulted in the rather high values of elementary rate of bed load transport in the S1 and S2 measurement sites. In the second part of the measurement period, with stabilization of the valley sides, the Scott Glacier became the main source of material supply. Generally, a lower discharge and long distance of transport favoured selective erosion, resulting in the formation of a channel pavement. For that reason the critical velocity of flow necessary to initiate bed load movement was higher. This fact resulted in the diametrically different amounts of bed load transported in successive days, although the discharge rate was similar (Kociuba et al. [7]).

5 Conclusions

The results of investigations conducted in the research profile of the Scottelva in the summer season of 2009 indicated a great temporal and spatial variability of bed load transport, the amount of which in individual measurement sites ranged from zero to several dozen kilograms for 24 hours. The increased sediment load in the river water was accompanied by considerable changes of channel geometry. We registered more intensive down-cutting and processes of lateral and vertical accretion of deposits: formation of bars, gravel shadows, levees and building up of channel bed.

The investigations undertaken with the application of the new survey methodology indicated a great temporal and spatial variability of fluvial processes conditioned by hydro-climatic factors – particularly glacier condition and ablation rate, glacier extent in the catchment, thickness of the permafrost active layer and weather and thermal conditions – determining the discharge value.

References

[1] Bartoszewski S., *Reżim odpływu rzek Ziemi Wedel Jarlsberga (Spitsbergen)*. Wyd. UMCS, Lublin, pp. 1-167. 1998

[2] Bartoszewski S., Gluza A., Siwek K., Zagórski P., The functioning of Scott Glacier in conditions of climate global changes. *Landform Analysis*, 5, pp. 5–8. 2007

[3] Bunte K., Potyond J., P., Abt S., Sampler size and sampling time affect measured bedload transport rates and particle sizes measured with bedload traps in gravel-bed streams, [in]: *Erosion and Sediment Transport Measurement in Rivers: Technological and Methodological Advances*. J. Bogen, T. Fergus and D.E. Walling (eds.), IAHS-Publication No. 283, pp 126–133. 2003

[4] Chmiel S., Bartoszewski S., Gluza A., Siwek K., Zagórski P., Physicochemical characteristic of land waters in the Bellsund region (Spitsbergen). *Landform Analysis*, 5, pp. 11–13. 2007

[5] Froehlich W., Akustyczne i magnetyczne badania transportu ładunku dennego, [w:] Florek W., Kaczmarzyk J., (red.), Współczesne problemy geomorfologii, *Landfom Analysis*, 9, pp. 348–350. 2008

[6] Hammer K.M., Smith N.D., Sediment production and transport in proglacial stream: Hilda Glacier, Alberta, Canada. *Boreas*, 12, pp. 91-106. 1983

[7] Kociuba W., Janicki G., Siwek K., Variability of sediment transport in the Scottelva catchment in the hydrologically active season of 2009. *Questiones Geographica A*. in press

[8] Pękala K., Repelewska–Pękalowa J., Dynamics of active layer of permafrost. *Landform Analysis* 5, pp. 168–169. 2007

[9] Rachlewicz G., Floods in high Arctic Valley systems and their geomorphologic effects (examples from Billefjorden, Central Spitsbergen). *Landform Analysis*, 5, pp. 66–70. 2007
[10] Reder J., Zagórski P., Recession and development of marginal zone of the Scott Glacier, *Landform Analysis* 5, pp. 175–178. 2007
[11] Szczęsny R., Dzierżek J., Harasimiuk M., Nitychoruk J., Pękala K., Repelewska-Pękalowa J., Photogeological map of Renardbreen, Scottbreen and Blomlibreen Forefield (Wedel Jarlsberg Land, Spitsbergen), scale 1:10 000. Wydawnictwa Geologiczne, Warszawa. 1987
[12] Warburton J., An alpine proglacial fluvial sediment budget. *Geografiska Annales*, 72A, pp. 261–272. 1990
[13] Zagórski P., Siwek K., Gluza. A., Bartoszewski S., Changes in the extent and geometry of the Scott Glacier, Spitsbergen, *Polish Polar Research*, 29, 2, pp. 163–185. 2008

Author Index

Ballio F. 149
Berzi D. 41
Bisantino T. 199, 237
Bomont S. 113
Brambilla D. 149, 185
Bugnion L. 161

Cheng Z. L. 101
Comiti F. 225
Corrêa G. P. 77
Coussot P. 51

De Gonzalo Aranoa C. 213
De Wrachien D. 3
Deangeli C. 3
Di J. Y. 65

Evangelista A. 51

Fischer P. 237

Garcia-Martinez R. 29
Gardner J. S. 137
Gentile F. 199, 237
Giorgetti E. 149
Goerl R. F. 77

Huelin Rueda P. 213

Janicki G. 261
Jenkins J. T. 41
Jiang Y. 137

Kobiyama M. 77
Kociuba W. 261

Larcan E. 41
Lenzi M. A. 249
Li J. 65
Li Y. M. 65

Liu J. J. 101
Liu J. K. 101
Longoni L. 149, 185

Mambretti S. 3
Mao L. 225, 249
Martinez C. 29
Michel G. P. 77
Mintegui Aguirre J. A. 213, 249
Miralles-Wilhelm F. 29
Mrvík O. 113

Papini M. 149, 185
Pavan S. 17
Pellegrino A. M. 51

Radice A. 149
Riedl F. J. 125, 175
Robredo Sánchez J. C. 213, 249

Schippa L. 17
Scotto di Santolo A. 51
Siwek K. 261

Trisorio Liuzzi G. 199, 237

Wei F. 89, 137
Wendeler C. 161

Xu A. 137
Xu F. W. 65
Xu J. 65

Yang H. 89
Yang X. D. 65

Zhang G. P. 65
Zhao L. N. 65
Zhao Y. 89, 137